HOW TO RESTORE

Braking Systems

OSPREY

RESTORATION

GUIDE 10

HOW TO RESTORE

Braking Systems

Joss Joselyn

Published in 1985 by Osprey Publishing Limited
12–14 Long Acre, London WC2E 9LP
Member of the George Philip Group

Sole distributors for the USA

Publishers & Wholesalers Inc
Osceola, Wisconsin 54020, USA

British Library Cataloguing in Publication Data

Joselyn, Joss
　　How to restore braking systems.—(Osprey
　　restoration guide; 10)
　　1. Automobiles—Brakes—Maintenance and repair
　　I. Title
　　629.2′46　　TL269
　　ISBN 0–85045–645–2

Editor Tim Parker

Filmset by Tameside Filmsetting Limited
Ashton-under-Lyne, Lancashire
Printed by BAS Printers Limited
Over Wallop, Stockbridge, Hampshire

CONTENTS

Introduction

Brakes haven't altered fundamentally over the years but most of what change there has been took place in the postwar era. Cable-operated drums, front and rear—that's the set-up mostly fitted when manufacturers started to produce the old pre-war models again. Then cables gave way to hydraulics, drums were superseded largely by discs, on the front axle anyway, and servo assistance became a lot more common.

A fundamental similarity is apparent across the products of all the different manufacturers. One drum brake looks much like another, and there's only a limited number of variations in disc brake design. It's not necessary, therefore, to go into every manufacturer's products in enormous detail. Girling is featured more than many because that company's designs were most widely adopted by manufacturers in the fifties and sixties. Lockheed, in chapter 2, made the brakes for some very popular models, Bendix appeared a lot on French cars, as well as in this book's chapter 3, and chapter 4 contains representative coverage of American designs.

Because of the basic concept similarity, it's best to read all the first four chapters, irrespective of who made your car's braking system. Even if it's a make not mentioned in this volume, all the fundamental principles of overhaul will still apply.

Before committing yourself irrevocably to a comprehensive replacement programme, make sure the new parts are all available. In some cases, where the brake manufacturers can no longer supply, the owners' clubs can oblige. Many older lines were 'remaindered' to them by the manufacturers a few years back.

Where replacement brake shoes are no longer available, it is usually possible to get your old shoes re-lined by a service run by Ferodo Ltd. Contact them either at their headquarters or one of their main branches or agents.

Because of a desire for a car to be in 'original' condition, few restorers are likely to try to alter the original brake specification. Even if tempted, don't try it—one small change can badly upset the entire system and render it inefficient—and who wants a car with sub-standard brakes?

One last cautionary word; because gaining access to brakes usually means removing the road wheels, the car has to be supported clear of the ground. *Never* work on a vehicle supported only on a jack. Axle stands are very little more trouble but they are safe. Use them properly and chock the wheels left on the ground, and you'll find that most car's braking systems are safely within the scope of the amateur mechanic.

Finally, I would like to thank Lucas Girling Ltd, Automotive Products Plc, AC-Delco (General Motors), Renault UK Ltd and Talbot Motor Co. Ltd who gave us permission to reproduce a number of illustrations from their manuals.

Joss Joselyn
September 1985

Chapter 1 | **Girling brakes**

In the 25 years after the war Girling were one of the two major braking firms and shared with Lockheed the major slice of original equipment contracts. Although this was a period when motor car braking was gradually being improved, it pre-dated the years when frequent minor modifications made the whole scene complicated. The vast array of cars was in fact served by very few different brakes. The one innovation that was particularly notable, however, was the disc brake, particularly for the front wheels. Eventually every manufacturer made the change from drum to disc, but some took longer than others.

Front drum brakes
Between 1945 and 1949 brakes were mainly systems inherited from pre-war and some were even hydro-mechanical systems with hydraulic brakes on the front and cable-operated at the rear. One brake that was still in use up until 1949 was the non-servo brake. Although it was cable-operated and had been around since 1930, it can be seen now to have some features recognizable as Girling many years later (Fig. 1:1).

Another early drum brake, used on postwar Armstrong Siddeley Sapphire, Rover, Daimler, Humber Super Snipe and Jaguar, was known as the 'autostatic two trailing shoe.' Whereas most more modern brakes are of a twin-leading design, this one, shown in Fig. 1:2, has wheel cylinder pistons which operate against the forward rotation of the drum. Normally requiring a high input force, it is a lot less afflicted by brake fade when hot.

Many of its features appear in later versions, but perhaps

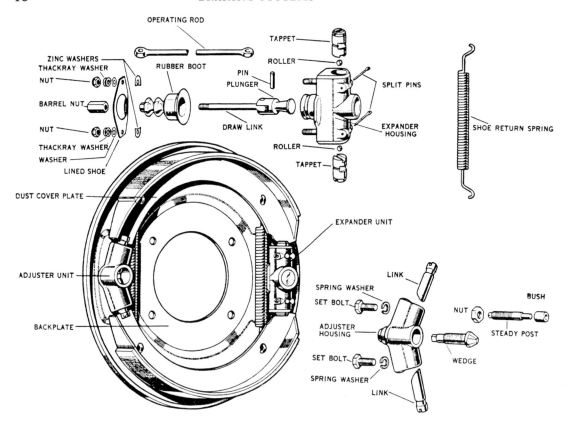

OPERATING ROD

TAPPET

ROLLER

ZINC WASHERS
THACKRAY WASHER
NUT
RUBBER BOOT
PIN
PLUNGER
SPLIT PINS
BARREL NUT
NUT
DRAW LINK
EXPANDER
HOUSING
THACKRAY WASHER
WASHER
LINED SHOE
ROLLER
TAPPET
SHOE RETURN SPRING

DUST COVER PLATE

EXPANDER UNIT

LINK
SPRING WASHER
SET BOLT
BUSH
NUT
ADJUSTER UNIT
ADJUSTER
HOUSING
STEADY POST
BACKPLATE
SET BOLT
WEDGE
SPRING WASHER
LINK

Fig. 1:1. Introduced in 1930 and first of a long line, this is the Girling non-servo brake

the only one that does not is the stabilizer (Fig. 1:3) which occupies the position of the hold-down spring in later designs. This was, in fact, a primitive form of adjuster. The shoe web is slotted and while movement of the brake overcomes the grip of the friction washer, the pull of the return spring does not. Hence the shoes are retained in close proximity to the drum.

The design that was popularly used from about 1953, particularly by Ford and Vauxhall, but also by Morris on the Oxford and 1.3 Marina and by Hillman for the Imp, was the twin-leading shoe H.L.S.S. (hydraulic leading shoe, sliding). All this means is that each brake shoe is operated by its own wheel cylinder and the ends of the shoes are designed to slide in the abutments in the wheel cylinders to provide more efficient use of the whole lining.

Adjustment is by means of two snail cams—one for each shoe. See Fig. 1:4.

Assuming that nothing is known of the car's braking system and a complete overhaul is envisaged, you will obviously be prepared to change parts and also realise that it may not be just a simple couple of hours' job. Support the front end of the car on axle stands; don't leave it on a jack, and if you can beg, borrow or steal a Girling service tool set, or even buy one from your local main Girling agent, it will be found very useful.

Fig. 1:2. Another early Girling brake, used on a number of different makes, is this autostatic two trailing shoe design

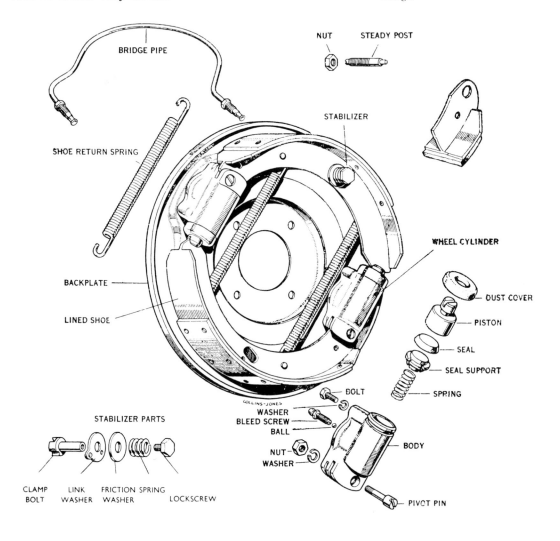

NUT   STEADY POST

BRIDGE PIPE

STABILIZER

SHOE RETURN SPRING

WHEEL CYLINDER

DUST COVER

PISTON

BACKPLATE

SEAL

LINED SHOE

SEAL SUPPORT

SPRING

COLLINS-JONES

BOLT

WASHER
BLEED SCREW
BALL

BODY

NUT
WASHER

STABILIZER PARTS

PIVOT PIN

CLAMP   LINK   FRICTION SPRING
BOLT   WASHER   WASHER   LOCKSCREW

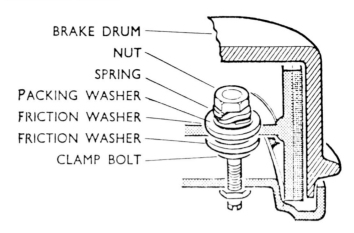

BRAKE DRUM
NUT
SPRING
PACKING WASHER
FRICTION WASHER
FRICTION WASHER
CLAMP BOLT

Fig. 1:3. The stabilizer in the autostatic two trailing shoe brake. It was really a type of adjuster

With the car securely supported, remove the road wheels, back off the adjusters and take off the drums. Tackle the hold-down springs in the centre of each brake shoe web next; pliers may be used to push down on the spring and twist the flattened end of the post to allow it to pass through the spring plate. Be careful not to allow the bits to fly all over the workshop and note that, with the coil type hold-down springs, the Girling tool (64947090) makes the job a lot easier.

Another special tool—private car shoe horn (64947019)—is recommended for removing the shoes against shoe return spring pressure. A good, strong screwdriver can be used instead to lever the ends of the shoes out of their abutments, but be careful not to lever against the dust-cover ends of the wheels' cylinders or they can be damaged. Normally, the recommendation is made that the position of the shoe return springs is noted so they can be re-fitted correctly, but as they could have been wrongly installed in the first place by a previous owner, it is safer to refer to the diagram.

Check that each snail cam adjuster head is square and undamaged and then use a spanner to turn it. It should not be seized but should not be totally free either. If there's no resistance at all, the cam will work off through shoe return spring pressure and result in long pedal action. If the adjusters are seized, rotate them to and fro to free them (no penetrating oil). If the action is satisfactory, clean the outside of the backplate and put a smear of Girling brake grease at the point of movement.

If the adjuster is loose or won't free off, a replacement will have to be fitted, using an adjuster Service Kit. Note the position of the old adjuster and then gently file away the riveted head and remove it. Don't use excessive force or you'll damage the backplate. Smear the new adjuster stem with brake grease and fit it. Hold it firm with a brake spanner, slip the new spring washer over, fit the new cam in the position of the old one, and screw on the locking nut tightly (Fig. 1:5).

Have a look at the wheel cylinders to see if they are leaking. The simplest way is to hook back the lip of the dust

Fig. 1:4. The Girling H.L.S.S. (hydraulic leading shoe, sliding) brake. A popular design used by Ford, Vauxhall, Morris and Hillman

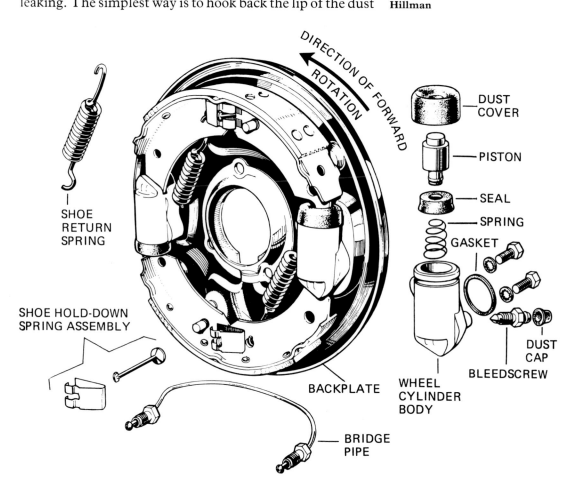

DIRECTION OF FORWARD ROTATION

SHOE RETURN SPRING

SHOE HOLD-DOWN SPRING ASSEMBLY

BACKPLATE

BRIDGE PIPE

WHEEL CYLINDER BODY

DUST COVER

PISTON

SEAL

SPRING

GASKET

DUST CAP

BLEEDSCREW

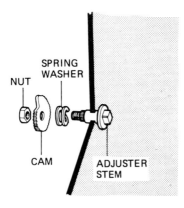

SPRING
WASHER

NUT

CAM   ADJUSTER
                STEM

Fig. 1:5. An exploded view of the
parts in a snail cam adjuster. If
it is seized or too loose, it must
be replaced using an adjuster kit

covers; if there is fluid escaping, new cylinders should be
fitted on both sides. Try the pistons to ensure they are
moving freely, and if all is well, secure them with elastic
bands to ensure they aren't accidentally shifted.

Wash down the backplate with Girling cleaning fluid.
It's recommended because it's safe and petrol and paraffin
most definitely are not. Any corrosion can be removed with
a wire brush, but make sure you don't damage the rubber
dust covers on the cylinders. Locate the platforms on the
backplate against which the brake shoes move and smooth
them off with emery cloth. Wipe them clean and apply a
thin smear of brake grease, keeping it well away from both
wheel cylinders and brake shoe linings.

Use emery cloth to ensure the tips of the new shoes are
clean and smooth and apply a smear of brake grease.
Remove the elastic bands from cylinders and fit the shoes.
Attach the new shoe return springs in the appropriate holes
and fit the shoes into the piston (rubber dust cover) ends of
the cylinders first; then lever the other ends against spring
pressure into the abutments. Ensure that the shoes go in
the right way round, i.e. relating to the direction of
rotation, keeping the longest unlined part of the shoe at the
back. The shoe lining should look as though it has been
displaced on the shoe in the forward direction of rotation.

Wipe the drum clean ready for refitting. Do not blow out
the dust with an airline; asbestos is dangerous when
inhaled.

Fig. 1:6. The steady posts hold
the shoes square with the drum.
Where they are adjustable, the
following is the method:

A   Loosen locknut.
B   Unscrew (screwdriver in
     the slot) the adjustable
     steady posts two turns
     anti-clockwise.
C   Adjust shoes to the drum
     by the normal shoe
     adjusters until the drum
     cannot be turned.
D   Turn the steady post
     clockwise until it just
     contacts the shoe. Light
     contact only is required.
E   Tighten locknut.
F   Adjust shoes by releasing
     the brake adjuster to give
     correct clearance between
     shoe and drum.

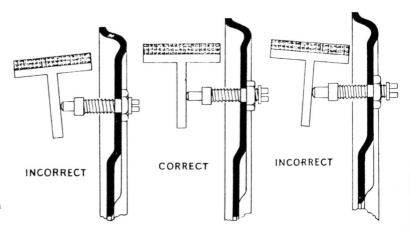

INCORRECT   CORRECT   INCORRECT

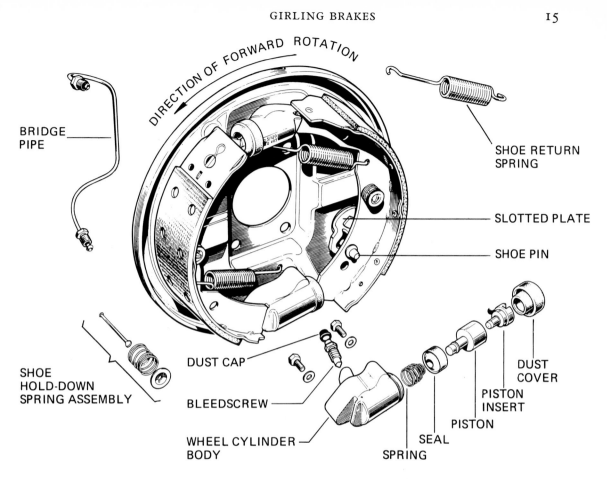

DIRECTION OF FORWARD ROTATION

BRIDGE PIPE

SHOE RETURN SPRING

SLOTTED PLATE

SHOE PIN

SHOE HOLD-DOWN SPRING ASSEMBLY

DUST CAP

BLEEDSCREW

WHEEL CYLINDER BODY

SPRING

SEAL

PISTON

PISTON INSERT

DUST COVER

If either leaf type or coil type hold-down springs are fitted, re-install these. If, as were used on some older brakes, there are adjustable steady posts (Fig. 1:6), these will have to be reset. Fit the drum first, then slacken off the locknuts on the backplate, and screw the steady posts outwards for two full turns. Lock the shoes hard in the drum by tightening the adjusters clockwise, then screw in each steady post until it makes contact with the shoe web. Then tighten the locknuts, ensuring at the same time that the steady posts do not move. Adjust the brake by slackening off the adjusters just enough to allow the drums to move without dragging.

Repeat the overhaul sequence on the other side. Remember, anything that is changed should be done

Fig. 1:7. The H.L.S.A.S. (hydraulic leading shoe auto-adjust by service brake). This is similar to the previous brake but with automatic adjustment

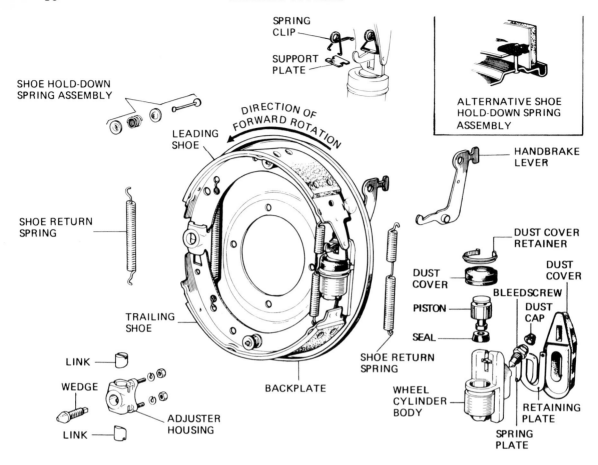

SPRING CLIP

SUPPORT PLATE

ALTERNATIVE SHOE HOLD-DOWN SPRING ASSEMBLY

SHOE HOLD-DOWN SPRING ASSEMBLY

DIRECTION OF FORWARD ROTATION

LEADING SHOE

HANDBRAKE LEVER

SHOE RETURN SPRING

DUST COVER RETAINER

DUST COVER

DUST COVER

BLEEDSCREW

DUST CAP

PISTON

SEAL

TRAILING SHOE

SHOE RETURN SPRING

LINK

WEDGE

BACKPLATE

WHEEL CYLINDER BODY

RETAINING PLATE

LINK

ADJUSTER HOUSING

SPRING PLATE

Fig. 1:8. The H.L.3 (hydraulic lever Mk 3) brake. The main difference between this rear brake and the foregoing front ones is that this has leading and trailing brake shoe arrangement

'across the axle'. If new shoes are fitted, install them on both sides, and the same with new wheel cylinders.

One final note. The makers always recommend that new shoe return springs are fitted. In practice this is not done in every case, but if the existing springs are of totally unknown age and condition, new ones could be a sound move.

The only variation of this popular brake was the addition of a simple automatic adjustment device to replace the snail cams. This was the H.L.S.A.S. (hydraulic leading shoe auto-adjust by service brake) and it was fitted to the Vauxhall Victor FC. It consisted of an actuating pin on the

Left **Fig. 1:9.** It is important with this brake to ensure the cylinder is free to slide. Grease is applied using a feeler blade

Below **Fig. 1:10.** The H.L.3A brake. The main difference between this and the H.L.3 is the automatic adjustment operated by movement of the handbrake

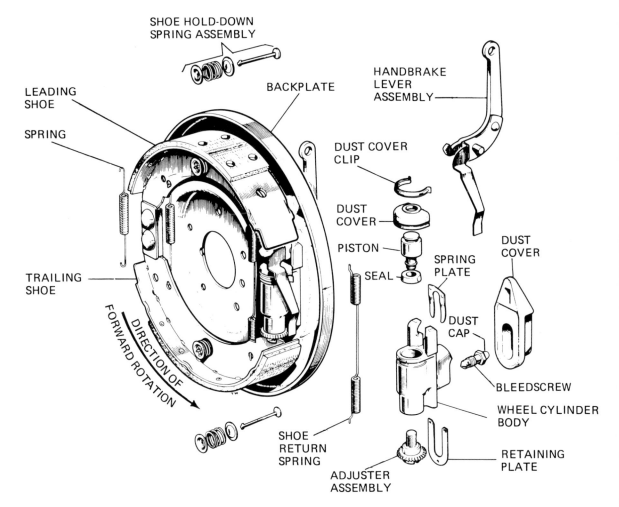

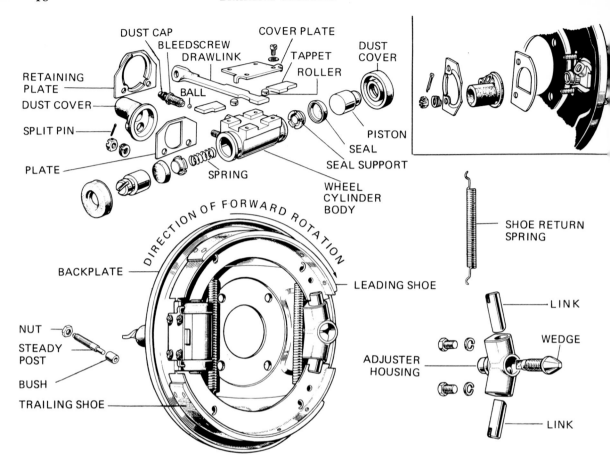

Fig. 1:11. This is the H.W. brake (hydraulic wedge). The main point of difference here is the separate mechanical expander for the handbrake. There is also another version of this which has not been illustrated

web of the shoe which engages in a slotted plate attached to the back plate via a friction washer (Fig. 1:7). Correct shoe-to-drum clearance is maintained by a clearance between the pin and the locating slot in the plate. When the foot brake moves the shoes, the clearance between pin and slot is taken up and any further movement moves the friction-loaded plate. The shoe return springs are not strong enough to move the plate, so when the shoes are pulled back off, the plate stays put. The only direction it will move in is further towards the drum to take up further lining wear.

The only extra work involved is to adjust the friction loaded plate initially when fitting new shoes.

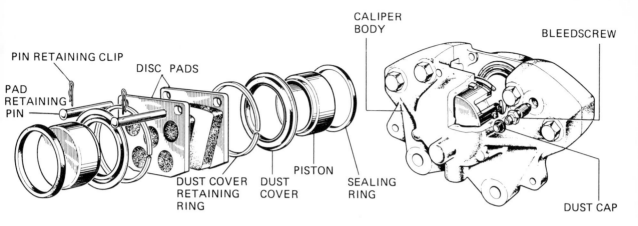

Rear drum brakes

Much of the information on front drum brakes applies equally to rear drums brakes, and although three different main Girling types have been used, they are essentially similar. The principal differences between front and rear are that rear brakes are single leading and trailing shoe—i.e. they have one double-acting cylinder—and have a different adjusting mechanism; some manual and others automatic.

The brake most commonly encountered is the H.L.3 (hydraulic lever Mk 3). See Fig. 1:8. Additional points to watch when overhauling are first to inspect, clean and grease the wedge adjuster. Any parts that are damaged should be renewed and it's particularly important to ensure that the threaded stem will run its full length and leave the adjuster in the fully retracted position. Use Girling brake grease as the lubricant.

The other additional feature is the attachment of the handbrake linkage. Where a support plate and spring are fitted to the leading shoe, they should be retained to protect the shoe from the action of the hardened handbrake lever tip.

Undoubtedly the most important part of any overhaul on this brake is to ensure that the wheel cylinder is able to slide freely in its slot in the backplate. It frequently seizes and the way to free it is usually to tap it gently to get it moving, and then to work brake grease between cylinder

Fig. 1:12. This is the most commonly used Girling disc brake. It used horizontally opposed pistons and in some cases there were four of them

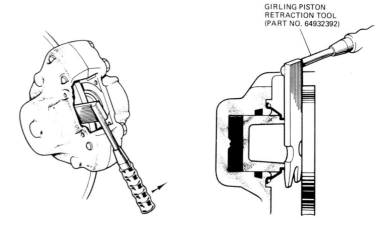

GIRLING PISTON
RETRACTION TOOL
(PART NO. 64932392)

Fig. 1:13. Using the Girling tool to lever back the pistons to make space for the new pads

and backplate using a feeler gauge (Fig. 1:9). It may be necessary to remove it to clean out corrosion, but whatever it takes, it must be free or partial braking and uneven lining wear will result.

Developed from the H.L.3 was the H.L.3A which is similar except that it has an automatic adjustment mechanism (Fig. 1:10). There has been a lot of controversy over this mechanism in the past. Many people maintain it has never worked satisfactorily. Girling say that provided it is properly maintained and adjusted, there is no problem.

Check first that the wheel cylinder slides on the backplate. Then ensure that the ratchet wheel is not worn, turns easily on the adjustment screw and will travel freely the full length of the thread. Clean the threads and smear with brake grease.

When reassembled, turn the adjuster ratchet wheel until it is just possible to slide the brake drum over the shoes. Do the same thing on the other side. Operating the handbrake 20 or 30 times should achieve correct adjustment.

Remember that correct handbrake lever operation on an automatically adjusting system is six or seven clicks; not the normal three or four of a manually adjustable system.

The third commonly used Girling system is the HW (hydraulic wedge) a generally larger assembly for use on more powerful cars (Fig. 1:11), the main mechanical variation in this brake is the separate mechanical expander for handbrake operation. There are two versions which are similar acting, although slightly different mechanically. The essential check is that the wheel cylinder can move on the backplate in the HW version where the handbrake drawlink and mechanism is housed in the wheel cylinder body; and for the mechanism components to move on the wheel cylinder in the H.W2. Essentially all the parts should be clean, lubricated and in good condition. To overhaul completely or change parts, the wheel cylinder must be removed and this is covered in detail in chapter 5.

There were a number of other handbrake operated automatic adjustment mechanisms, many developed for individual cars and individual handbrake mechanisms. All of them, prior to the later development when they became service brake operated, used a small lever to operate a ratchet adjustment which moves on a threaded rod to expand the shoes to take up wear. A clean and lubricated thread, an unworn ratchet, and free but not worn pivot points are what to look for when overhauling. Correct

Fig. 1:14. There are differences on this Girling-Dunlop Mk 2 caliper. It was less used than Girling's own, but could be encountered

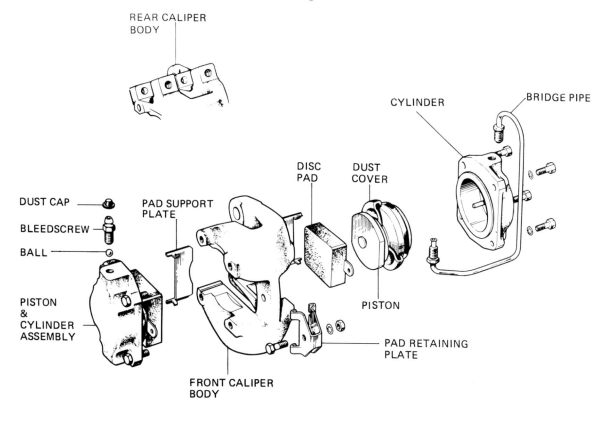

handbrake adjustment is also important but that is dealt with in chapter nine.

Disc front brakes

Probably the era's greatest braking advance, discs began to come into general use during the 1950s. The Girling caliper used on a wide variety of cars of the day had horizontally opposed pistons (Fig. 1:12). The number of pistons varied between two and four and there were different sizes, but the general principle of this unit was the same throughout the range. Never unbolt the two caliper halves.

An overhaul, whether motivated by some specific fault or carried out as preventive maintenance, would start by fitting new pads. This is an operation probably familiar to most people, but the following is a brief summary.

Ensure the car is firmly supported on axle stands and remove the front wheels. Clean up the caliper with a wire brush and note the position of any damping shims.

Pull out the little wire clips, followed by the retaining pins and then the actual pads; pliers will usually accomplish this. Look at the pads as they come out and excessive wear on one of the pair may indicate a 'lazy' piston or one that has seized, but the work to deal with this is gone into in detail in chapter five.

A bit of cleaning up is important at this stage. Tackle the disc first, and use a screwdriver held against the corrosion build-up outside the swept area, while spinning the disc, to scrape it off. Finish with emery cloth. Use this too to clean off any corrosion in the pad apertures, especially the areas where the new pads seat.

The pistons need to be retracted into their cylinders next. To do this either undo the bleed nipple a turn and then use the Girling tool to lever them back (Fig. 1:13), or remove the master cylinder cover and wrap old rags around it to collect the fluid that overspills. If you don't have the special retraction tool, you'll have to make do with something else, but don't tilt the pistons. A piece of hardwood is probably best because it won't damage them.

Any damping shims should be replaced. Clean them first and coat them both sides with the special squeal-deterrent

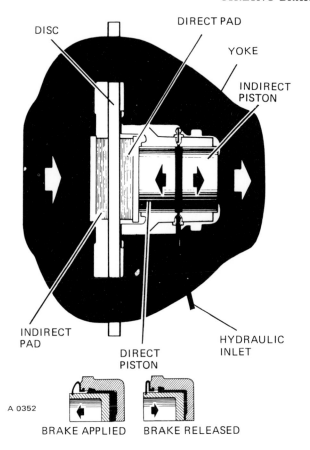

DISC

DIRECT PAD

YOKE

INDIRECT PISTON

INDIRECT PAD

DIRECT PISTON

HYDRAULIC INLET

A 0352

BRAKE APPLIED   BRAKE RELEASED

Fig. 1:15. The Girling 'A' type caliper has both pistons on one side of the disc. It was used on the Austin Maxi and Allegro as well as others

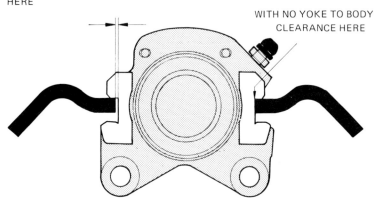

YOKE TO BODY CLEARANCE TO BE 0.006 IN. (0.15 MM) TO 0.012 IN. (0.30 MM) HERE

WITH NO YOKE TO BODY CLEARANCE HERE

Fig. 1:16. It is important that the mating surfaces of yoke and cylinder are free from corrosion. The amount of clearance is also important

grease supplied with the shoes. Coat the backplates of the new pads, taking great care not to get grease on the friction material. Reassemble and fit new clips if the old ones look dodgy. Repeat the procedure on the other brake and finally pump the brake pedal to bring the pads back in contact with the disc. Top up the master cylinder level and road test.

Used far less than Girling's own caliper were a couple of Girling/Dunlop types. These are quite a lot different. The piston/cylinder assemblies, for instance, are bolted to the outside of the caliper saddle; The pads are directly connected to the pistons; the pistons have a mechanical retraction device rather than relying on the distortion resistance of the dust covers; and the pistons' seals which are static in the cylinder walls in Girling calipers, are mobile on the piston in the Dunlop.

Other differences occur when changing pads. On the Girling Dunlop Mk. 2 (Fig. 1:14), a nut and bolt and a retaining plate are removed to release the pads. On the Series 3 it is a single long retaining pin. On both types the raised spigot on the pistons which locate the pads must be checked for damage. All the other points—cleaning rust off the discs, checking seals, retracting the pistons are all similar to the Girling caliper.

There was one other Girling front disc that was based on a different principle. Used in the Maxi and the Allegro, the 'A' type caliper consists of a yoke, cylinder assembly and pads. The pistons are both on one side of the disc and expand outwards (Fig. 1:15). One acts directly on one of the pads and the other on the yoke to slide it in cylinder grooves and bring the other pad to bear.

When overhauling, there are obvious differences of design, but the broad principles are the same. The main additional overhaul job is to clean up the sliding edges of the yoke and the grooves in which they slide in the cylinder body. Use a wire brush but do not remove more metal than necessary; a maximum gap of between 0.006 in. and 0.012 in. (0.015 in. top whack) is recommended (Fig. 1:16).

Disc rear brakes

Logically enough, one of the two Girling, rear disc brakes

Fig. 1:17. A typical view of the Girling disc brake in service. This is the most commonly used type with horizontally opposed pistons

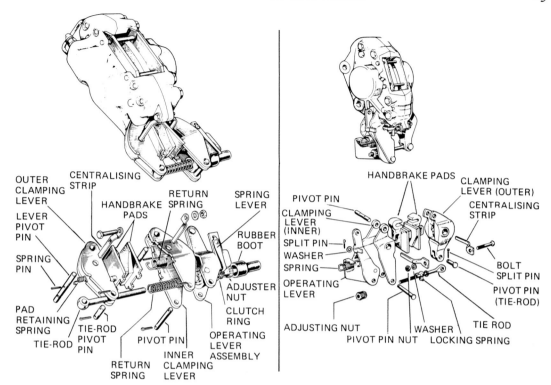

OUTER
CLAMPING
LEVER

CENTRALISING
STRIP

HANDBRAKE
PADS

RETURN
SPRING

SPRING
LEVER

SPRING
LEVER

LEVER
PIVOT
PIN

RUBBER
BOOT

SPRING
PIN

ADJUSTER
NUT

PAD
RETAINING
SPRING

CLUTCH
RING

TIE-ROD

TIE-ROD
PIVOT
PIN

PIVOT PIN

OPERATING
LEVER
ASSEMBLY

RETURN
SPRING

INNER
CLAMPING
LEVER

HANDBRAKE PADS

CLAMPING
LEVER (OUTER)

PIVOT PIN

CENTRALISING
STRIP

CLAMPING
LEVER
(INNER)

SPLIT PIN

WASHER

BOLT

SPRING

SPLIT PIN

OPERATING
LEVER

PIVOT PIN
(TIE-ROD)

TIE ROD

ADJUSTING NUT

PIVOT PIN NUT

WASHER

LOCKING SPRING

mostly used is similar to the popular front disc brake design, but with a handbrake mechanism added. Maintenance and overhaul of the actual disc brake is no different from the front version; only the handbrake mechanism adds any complication. See Fig. 1:18. It involves taking off the centralizing strips, withdrawing the pivot pin from the tie rod and swinging the two levers apart before the pads can be removed. Maintenance is mainly a matter of ensuring the adjuster threads are clean and lubricated and that similar conditions apply to the pivot pins.

There is also another rear disc brake based on Girling's other front disc design, the 'A' type sliding caliper. This was only fitted to the Peugeot 504, but once again the service requirements of the actual disc are the same. The third rear disc brake is the S.1H, which is a single-sided swinging design. As this was fitted only to the Rover 2000/3500 range and Zephyr/Zodiac, there is only space simply to mention it here.

Fig. 1:18. The only added complication to a rear disc brake is the handbrake mechanism. Two types are shown here, but the main requirement is that they should be clean and lubricated

Chapter 2 | **Lockheed brakes**

Radical new concepts have never been a feature of car braking and it is not surprising, therefore, the Lockheed designs, in all their major aspects, look remarkably similar to those in the previous chapter by Girling. There are some differences, however, and it's worth looking at these in detail.

Front drums

On cars like the Minor 1000, A35, A40 and the Mini a simple twin-leading shoe design was used. Even earlier than that a leading and trailing set-up was installed for instance on the Mini, but this is similar in detail to the system used on the rear of many of these same cars.

If you've just acquired one of the models that used the popular twin-leading shoe design, a check and overhaul will probably have early priority. Work starts by jacking up the front end and supporting it firmly on axle stands. Access to the brakes is in the usual fashion—removing the road wheels, backing off the adjusters and pulling off the brake drums.

What you will find is seen in Fig. 2:1. The Micram adjusters are typical of the era, but there is also the possibility you'll encounter an eccentric pin adjuster fitted instead, and in some cases little tie springs fitted on the head of the pistons and to a pin in the web of the brake shoe (Fig. 2:2).

These springs will have to be removed first, before the shoes can be levered off against return spring tension. The Girling 'shoe horn' tool mentioned in the last chapter is useful here, but in its absence, do the job with a stout

screwdriver, taking care not to lever against the
piston/dust-seal end of the cylinders. They are easily
damaged and it's better to lift the shoes out of that end after
return spring tension has been removed.

Before actually removing the shoes it is often better to
note the position of the return springs and the correct way
round of the brake shoes (linings at the front end, bearing
in mind the direction of wheel rotation). Alternatively, if
you tackle the two backplates one at a time, you'll always
have the assembled one to refer to (bearing in mind they are

Below **Fig. 2:1. This is the
Lockheed twin leading shoe
drum brake used on the Morris
Minor 1000, Austin A.35 and
A.40 and most notably on the
early Mini. Note also the
alternative 'eccentric' type of
adjuster**

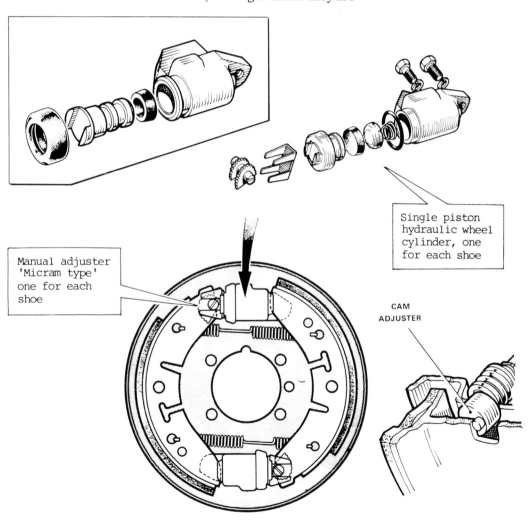

Single piston
hydraulic wheel
cylinder, one
for each shoe

Manual adjuster
'Micram type'
one for each
shoe

CAM
ADJUSTER

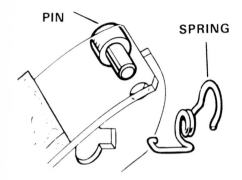

PIN   SPRING

Above **Fig. 2:2. Some versions of the Lockheed front drum brake had these little tie springs fitted between pistons and brake shoe webs**

Right **Fig. 2:3. This is the most commonly used Lockheed rear drum brake; it's a leading and trailing design and was used on the Mini**

handed), though even this does not allow for the possibility that the brakes have been wrongly assembled before. Our diagram (Fig. 2:1) is probably the safest reference.

While removing the shoes and springs, it will be necessary to free the shoes from the projecting pegs of the eccentric pin adjusters where these are used. Where Micram adjusters are fitted, these will have to be retrieved; they will fall clear as the shoes are removed.

Clean up the backplate initially with a wire brush and then using a cleaning fluid; ethyl alcohol (industrial methylated spirits) is what Lockheed recommend. Check the rubber dust covers on the cylinders, and 'lip up' the edge to check for fluid leaks. Any defects here will probably mean a new cylinder or perhaps rebuilding the existing one with a kit, but this is dealt with in chapter five.

Linings that have worn to less than one third of their original thickness should be renewed, say Lockheed, but if you're doing a thorough overhaul and you're not sure what linings might have been fitted in the past, a new set of genuine Lockheed shoes could be a good idea. Similarly, a new set of return springs would be playing it safe. Lockheed recommend comparing the existing springs with new ones to see if they appear weak or overstretched.

Wherever there is metal-to-metal contact, use a smear of brake grease, paying particular attention to the pivots on the Micram adjusters. This also means the tips of the new shoes, but taking care to keep the grease away from the rubber dust covers of the wheel cylinders. It must also be kept clear of the friction materials of the shoes and the surface of the drums. Where eccentric cam adjusters are used, simply ensure they are in good condition, not seized, and working properly (see Fig. 2:1).

Transfer the piston tie springs from the old shoes to the new and then fit them by reversing the removal procedure. The Micram adjusters, where fitted, go into the slots in the end of the shoe webs and the flat ends of the adjuster masks fit within the projections of the piston dust covers. These ends of the shoes are located first, then, with the return springs hooked in place, the other ends are levered into the abutments on the back of the cylinders. One advantage of

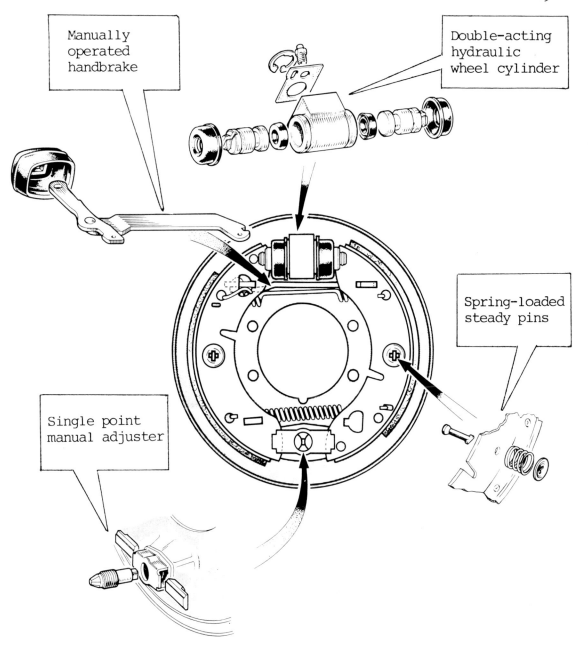

Manually operated handbrake

Double-acting hydraulic wheel cylinder

Spring-loaded steady pins

Single point manual adjuster

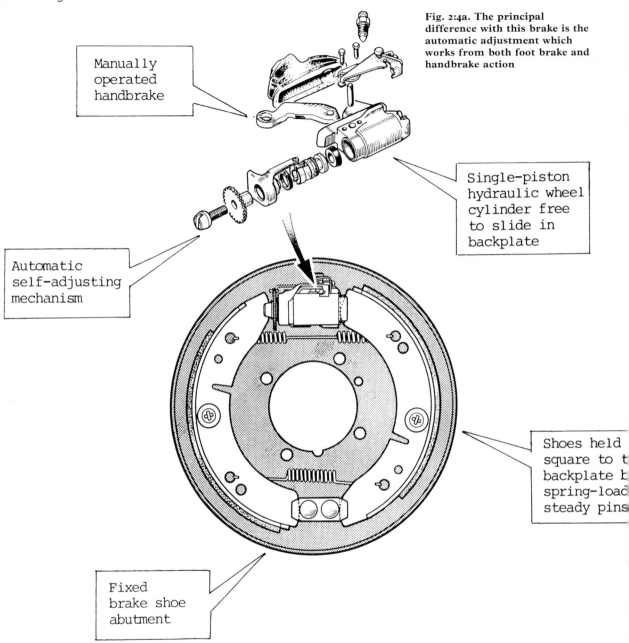

Fig. 2:4a. The principal
difference with this brake is the
automatic adjustment which
works from both foot brake and
handbrake action

Manually
operated
handbrake

Single-piston
hydraulic wheel
cylinder free
to slide in
backplate

Automatic
self-adjusting
mechanism

Shoes held
square to t
backplate b
spring-load
steady pins

Fixed
brake shoe
abutment

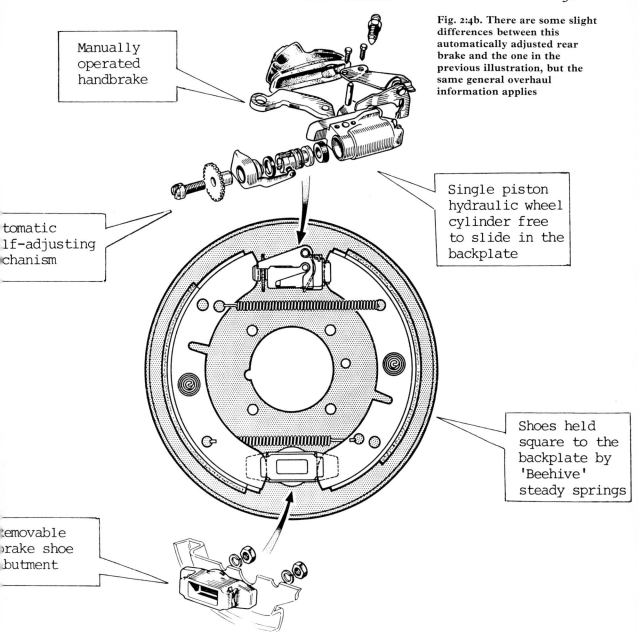

Manually operated handbrake

Fig. 2:4b. There are some slight differences between this automatically adjusted rear brake and the one in the previous illustration, but the same general overhaul information applies

Single piston hydraulic wheel cylinder free to slide in the backplate

tomatic lf-adjusting chanism

Shoes held square to the backplate by 'Beehive' steady springs

emovable rake shoe butment

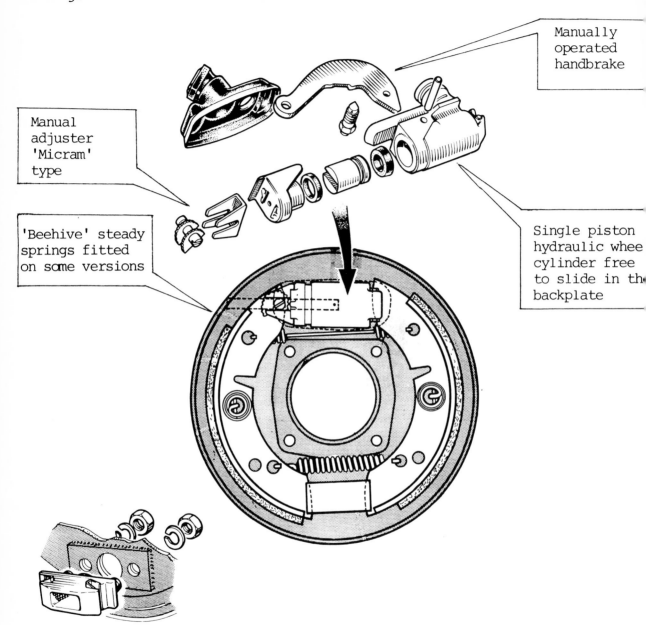

Manually operated handbrake

Manual adjuster 'Micram' type

Single piston hydraulic wheel cylinder free to slide in the backplate

'Beehive' steady springs fitted on some versions

Fig. 2:5. This design features a single acting wheel cylinder which centres itself by sliding in the backplate

this type of adjuster is that it is easy to renew. Before replacing the drum, ensure the adjusters are backed right off to the minimum setting, whether it is the Micram type or the less often seen eccentric cam. Clean the inside of the drum with cleaning fluid and check that the contact surface is not scored or damaged, and then refit and adjust.

Rear drum brakes

General principles are similar to Lockheed front brakes and again to Girling designs. There are three types, the first and most commonly used being a close relative to the front one already described but having only one double-acting wheel cylinder and a single point manual adjuster. This was also used on early Minis. The second has the addition of automatic adjustment and the third is a single-piston cylinder which is free to slide on the backplate, giving a leading and trailing action in both directions and a nice efficient handbrake. This type was used on the Hillman Minx and Minor 1000. All three obviously have a handbrake link.

The first, shown in Fig. 2:3, is tackled for overhaul purposes in much the same way as the others. This brake has hold-down or steady pins in the centre of the shoe webs, again similar to the Girling type with little coil springs, flat plates and a centre post with a flattened T-shape on the end, which can be twisted at right angles to lock the assembly. The Girling tool already mentioned (64947019) can be used with advantage here, or a pair of pliers if it is not available.

The only additional operation in removing the shoes is to disentangle them from the handbrake cross lever. There is an extra job to do on the backplate and that is to clean and lubricate the adjuster. Use a precise fitting brake spanner to ensure the thread is free along its entire length. Lockheed recommend PBC Shell corrosion resistant grease SB.2628 or an equivalent. This part of the operation is important because the threads on these adjusters are notorious for seizing.

Apply a thin smear of grease to the metal tips of the shoes, the contact points on the backplate and the handbrake cross lever pivot, check the wheel cylinder for

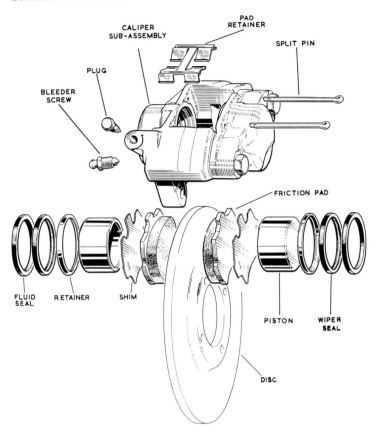

Fig. 2:6. This Lockheed disc brake design used horizontally opposed pistons, and is not dissimilar to the Girling type

leaks, etc., and remember if you fit new linings and return springs, do it on both rear brakes.

The principal difference in the second of Lockheed's rear drum brakes lies in the self-adjusting mechanism (Fig. 2:4). It works off both the foot brake and the hand brake, and like all these mechanisms, has to be carefully maintained in order to function.

The Lockheed overhaul procedure for this recommends removing the wheel cylinder and all the adjustment mechanism attached. Follow the procedure already described initially to remove the drum and the brake shoes, then disconnect the fluid inlet pipe from the wheel cylinder and take out the bleed screw. Disconnect the handbrake lever and ease off the rubber gaiter from cylinder and handbrake lever. Free the lug on the body near the bleed

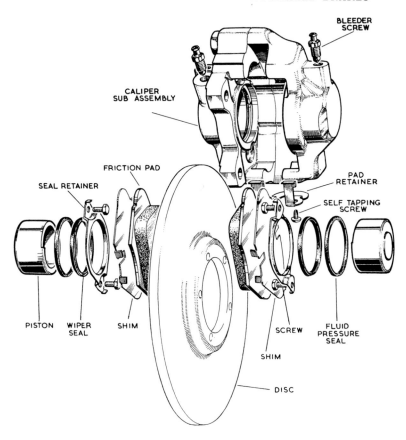

screw port and slide the cylinder assembly clear of its slot in the backplate.

On the bench, pull the adjuster wheel and screw out from the outer piston and dust cover. Take out two self-tapping screws to free the adjuster lever bracket from the cylinder body. Now take out the outer cylinder/dust cover.

Tap out the pivot pin to release the handbrake lever. The adjustment mechanism is now totally dismantled and can be cleaned up, checked and reassembled. Logically, however, having gone this far with dismantling the wheel cylinder itself, it would make sense to overhaul this at the same time by cleaning, checking cylinder condition, and renewing the rubber seals in the inner and outer pistons. Read the much more detailed description of what's involved in dealing with hydraulic components in chapter five.

Fig. 2:7. This also is an opposed piston type but heavier duty than the previous one. There are also some design differences

The inner piston and seal should be lubricated only with new brake fluid, but Lockheed 'Rubberlube' should be used for the inner contact surfaces of dust cover and attached piston (outer piston). This can also be used to increase the protection of the rubber gaiter.

Use brake grease sparingly to lubricate the adjuster screw and the shank of the adjuster wheel. Use 'Rubberlube' on the handbrake lever pivot hole and when the lever is located in the slot in the inner piston, tap the pivot pin home.

With all the outer piston components cleaned up, assemble them and coat the mouth of the bore, outer piston, seal and inner surfaces of the dust cover with 'Rubberlube' and refit the assembly into the cylinder bore.

The pivot point on the dust cover arm should be lubricated when it is fitted into its recess in the adjuster lever bracket, which is then refitted to the cylinder body with the self-tapping screws.

Screw the adjuster right home and then back it out again two and a half turns. Note that some assemblies have left-hand threads, identified by a slotted head or a hexagonal one. Insert the adjuster wheel and screw, ensuring the adjuster lever engages the wheel.

Grease the cylinder slot in the backplate and ensure it can slide when it's refitted. Refit the handbrake lever gaiter and install the bleed screw hand tight. Reassemble the backplate, but do not reconnect the handbrake lever at this stage.

Turn the adjustment until the drum can just be refitted. Reconnect the brake fluid pipe and bleed the air out of the system (see chapter 11).

Take the drum off again and back the adjuster wheel off for one whole turn. By getting someone to operate the foot pedal slowly and methodically, the action of the automatic adjuster can be checked. Heavy-footed pedal operation could cause the wheel cylinder outer piston to be ejected, so take it easy. Once certain it is working, it can be readjusted by hand, the drum refitted and the foot brake used to adjust it fully and finally.

Once the same work has been completed on the other side, the handbrake linkage can be reconnected and the handbrake and footbrake tested on the road. If the automatic adjustment doesn't work on the test, it is likely that worn parts will be the cause and renewal the answer.

The third of Lockheed's rear drum brakes is a return to the simpler side of life. This leading and trailing design (Fig. 2:5) uses one single-acting cylinder which slides in the backplate. It is adjusted by a single Micram adjuster and the shoes pivot at the other end on an abutment which is removable. All the points necessary to overhaul this brake have already been covered.

Front disc brakes

Calipers operating on two slightly different principles are manufactured by Lockheed. First is the type with two opposing pistons, one each side of the disc, and there are actually two versions of this—light and heavy duty. Then, in addition, there is the swinging caliper used on the Mk 11 BLMC 1100 and 1300. A single piston operates one friction pad directly, while reaction pivots the swinging caliper to which the piston is attached, and brings the other pad into contact with the disc.

The general principles of an overhaul on both of these are little different, characterized only by the varying number of pistons, pad-securing methods, and the need to ensure the pivot point of the swinging variety is free to move.

For the opposing piston type (Fig. 2:6), overhaul starts in the usual way by jacking up the front of the car and supporting it firmly on axle stands. Then, with the road wheels removed easy access to the calipers is possible. If a pad retaining spring, or springs, are fitted, these will have to be depressed in order to pull out the split pins, and then pliers used to extract the pads and any shims fitted behind them. Check these latter for damage or corrosion, but if they are sound they can be re-used.

Clean up the pad recess with emery cloth, using it very lightly simply to shift any corrosion. Also clean up the exposed part of the pistons but for this use only clean brake fluid. Provided the piston surfaces are undamaged, give the faces and pad recesses a smear of Lockheed disc brake lubricant and press each piston back into the caliper bore

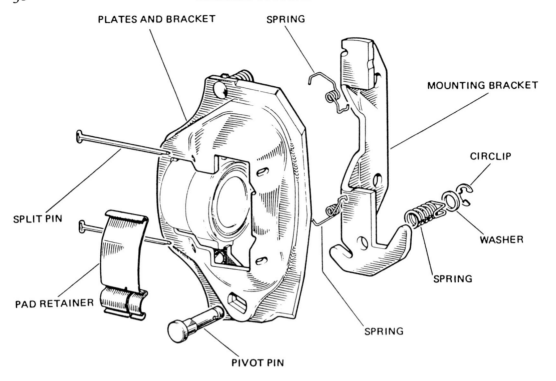

PLATES AND BRACKET

SPRING

MOUNTING BRACKET

CIRCLIP

SPLIT PIN

WASHER

PAD RETAINER

SPRING

SPRING

PIVOT PIN

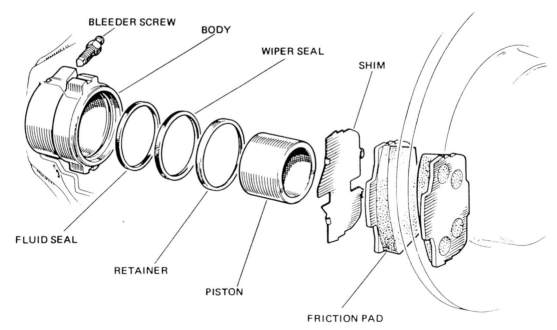

BLEEDER SCREW

BODY

WIPER SEAL

SHIM

FLUID SEAL

RETAINER

PISTON

FRICTION PAD

using the special Lockheed tool. This is a sort of clamp and if it is not available, it may be possible to use an ordinary G-clamp, if it will fit, or to retract the pistons using the Girling method of a lever that will not cause any damage.

Opening the bleed nipple is the simplest way to relieve pressure. Lockheed's own recommendation is to open the bleed nipple and use the foot pedal to eject some fluid. Then, with the master cylinder cap removed, there is room for the fluid level to move up the reservoir as the pistons are moved back.

Clean up the contact edges of the pad backing plates with emery cloth to remove any high spots. Lubricate them and the surface inside the pad window of the caliper with disc brake lubricant and fit pads, shims, and new retaining springs and split pins.

Clean any rust from the outer edge of the swept area on the discs. Finally, if the pistons show signs of damage, or there have been fluid leaks, further overhaul of the hydraulics will be needed, for which see chapter five.

With the heavy duty caliper (Fig. 2:7), there are some differences. First, the caliper has to be removed from the swivel axle (two bolts) in order to overhaul it, taking care to support it and not to strain the flexible brake hose. The pad retainer is also held by two screws, and with this gone, each pad is tilted within the caliper and withdrawn together with its shim.

The rest of the overhaul follows along standard lines, but care must be taken when refitting the caliper first to refit the shims originally removed, second to lock the bolts to the correct torque, and third to secure them by bending over the locking tabs.

With the swinging caliper (Fig. 2:8), bearing in mind that the two pistons are on one side of the caliper, there are no fundamental differences in overhaul. There is one slight additional job, which is to ensure that the pivot point on the caliper frame is free to move. If it's stuck a bit of scientific brutality with a soft-headed mallet should get it moving again. If it is too corroded, however, and cannot be freed off, a complete new caliper will be required.

Fig. 2:8. The Lockheed swinging caliper used notably on BLMC 1100 and 1300 models. Both pistons are on one side of the disc, and a free-moving pivot point is most important

Chapter 3 | **Bendix brakes**

Bendix brakes are mainly found on cars of Continental origin, particularly French. Renault, Peugeot, Citroen and Simca all used them. They were also fitted by Fiat and in later years by Ford, but once again they differ little in main principles from those of other manufacturers.

The procedure for overhauling drum brakes starts by jacking up the relevant end of the car and supporting it on axle stands. Drums may be secured either by two bolts as, for instance on the Simca 1000, or by removing grease cap, castellated nut and thrust washer as on, for instance, the Renault 16. Either way the adjustment should be backed off first, but this is not always as simple as it sounds, as the adjusters are exposed to the elements and often seize. Penetrating oil is the best bet to free them and use only the correct spanner to turn them as the squared adjusters are easily rounded off by a sloppy fitting spanner.

On the Simca, the drum securing bolts are screwed through threaded holes in the drum onto the hub to 'jack' it free; on the Renault a three-legged puller is supposed to be used, but it might be possible to do it by re-bolting the road wheel back on, turning it, and tapping the wheel rim with a hammer.

With the drums off, check them for scoring and damage and check the linings for uneven wear, which might mean, for instance, a sticking wheel cylinder.

In the case of rear brakes (Fig. 3:1), detach the handbrake cable (very similar brakes were fitted also to the front of early Simcas, as can be seen in Fig. 3:2).

The usual warning must be given here to make notes or draw diagrams, or something to ensure reassembly can be

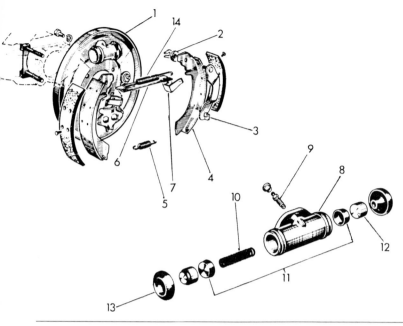

Fig. 3:1. Typical rear drum
brake assembly (on the Simca
1000)

1   Brake back plate
2   Handbrake lever pin
     retaining clip and washer
3   Handbrake lever
4   Rear brake shoe
5   Lower short spring
6   Handbrake lever pivot plate
7   Shoe anti-rattle clip
8   Wheel slave cylinder
9   Bleed screw
10 Seal cup centring spring
11 Seal cups
12 Piston
13 Dust cap
14 Upper long return spring

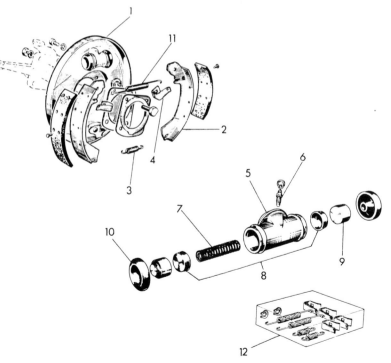

Fig. 3:2. Typical front drum
brake assembly (on the Simca
1000)

1   Brake back plate
2   Brake shoe
3   Lower short spring
4   Shoe anti-rattle clip
5   Wheel slave cylinder
6   Bleed screw
7   Seal cup centring spring
8   Seal cups
9   Piston
10 Dust cap
11 Upper long return spring
12 Spring and anti-rattle clip
     repair kit

done properly. Then a start can be made on dismantling, and the return spring (the one at the top) can be unhooked using a pair of pliers. The shoe steady clips are taken out next and the shoes pulled apart at the top so that the horizontal spacing strut can be taken out. Both shoes, still connected at the bottom by the return spring can then be lifted out as an assembly.

Lay them on the bench just as they were removed and use the assembly as a pattern for positioning the new shoes. Getting the leading and trailing aspect right is very important, so note carefully the position on the shoes of the friction linings, and which end the unlined metal areas are. When the shoes are correctly positioned, fit a new lower return spring between them. This is a very definite recommendation by the DBA people and they say that every time the brake shoes are renewed, new return springs should be fitted. Because they are in constant use and operate in a very hot environment, they eventually stretch and become less effective.

In the case of rear brakes, the handbrake levers will have to be retrieved and cleaned up ready for fitting to the new shoes. Before installing any of the new parts, clean off the backplate, using a wire brush and cleaning fluid. It's a good idea to wrap the wheel cylinder(s) around with an elastic band while doing this, but at one point remove the band and lift the dust cover to see if there are any signs of leaks. If there are or if one of the pistons is seized, the wheel cylinder will either have to be replaced or rebuilt (see chapter five).

Refitting all the components is generally the reverse of dismantling, but don't forget to put a thin smear of brake grease on the two ends of the brake shoes and on the handbrake lever pivot.

Bendix do produce an automatically adjusting rear drum brake, and although it is fitted to many current models, it is unlikely to be found on older cars selected for restoration. If anyone does want to overhaul this brake, the information on other types of auto-adjust in preceding chapters should allow the work to be carried out.

Bendix have fitted several different types of disc brake in the past but generally the most popular was a series of

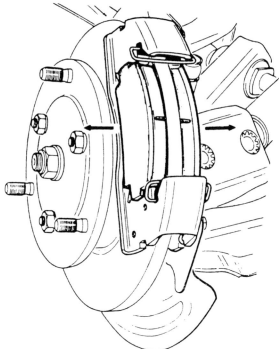

Above **Fig. 3:3. This is the Bendix III AS brake fitted to the Renault 16. The arrows indicate direction of pad removal**

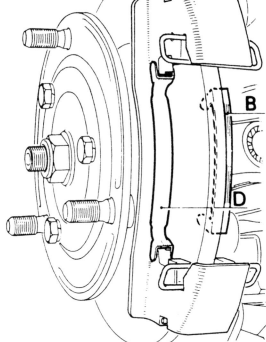

Right **Fig. 3:4. Again Bendix III AS, this time showing the location of the pad springs. Short springs on the inside (B), and long springs on the outside (D)**

single-piston sliding calipers, one version of which is still marketed today.

There is nothing extraordinarily different about a pad change and inspection of this brake, but nevertheless it is worth running through the sequence briefly.

With the car firmly raised and supported on axle stands, start by cleaning off the outside of the caliper unit. Soap and water can be used or some sort of non oily cleaning fluid, recommended by, or preferably made by, one of the brake manufacturers.

To remove the pads, first the securing clips are removed and this allows the little locking blocks to be extracted. It's best to use a small punch and mallet to tap out the first one, whereupon the second one should come out easily. This then allows the caliper assembly to be swung clear and supported. On the earlier Type 11 caliper this part of the operation is taking out two pins and releasing two swing clamps.

The pads come out next, upwards on Type II and sideways on Type III (Fig. 3:3). Note the position of the pad thrust springs before you remove them (Fig. 3:4).

Detach the rubber dust cover from its housing round the cylinder and use a piece of clean fluff-free cloth soaked in methylated spirits to clean the end of the piston. Use a block of wood or something similar to fill the cavity except for about 10 mm ($\frac{3}{8}$ in.) and use the brake pedal to eject the piston by that amount, but make sure the block prevents it from coming right out. Apply Spagraph grease to the protruding piston skirt and then use the block to push the piston back evenly into its cylinder. Before doing this, however, remove the top of the master cylinder reservoir and surround it with rags to absorb any brake fluid that overflows. When the piston is recessed again, refit the dust cover, cleaning it with methylated spirits first, or fit a new one. Insert the new pads.

Refit the pad thrust springs. You should have noted their positions before removing, but for the record, on Bendix II, the curved spring locates on the outside, the flat one on the inside (Fig. 3:5). With Bendix III, the long spring goes to the outside, short one on the inside (Fig. 3:4).

Finally, refit the caliper to the bracket and lock it with

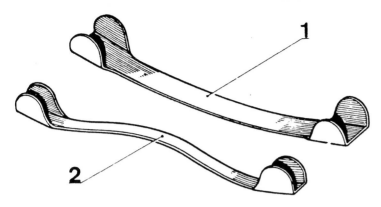

Fig. 3:5. Bendix II disc pad springs; 1 locates on the inside and 2 on the outside

the blocks and clips. Lower the car to the ground again and pump the brake pedal until strong resistance is felt (hard pedal), and you know the brakes are working. Never drive the car until this last operation has been carried out.

Chapter 4 | American brakes

To cover the whole spectrum of American automobile brakes would take a great deal more space than a single short chapter. It would also be somewhat boring, because although US cars are quite a lot different from those of Europe, the principles behind their brakes are not. They are still either disc or drum systems or a mixture of both. The units tend to be bigger and there are some detail differences, but covering a couple of typical units should be enough to give a good idea of what's involved in overhauling them.

Drum Brakes
From the vast array of slightly differing brakes fitted, the widely used Bendix has been selected as typical. It is used, for instance, by Ford, Chrysler and American Motors.

Inspection, shoe changing and general overhaul all start the same way—by jacking up the relevant end of the car, supporting it on axle stands and removing the road wheels. Before the drums can be pulled off, it will probably be necessary to back off the shoe adjustment. This is generally a matter of turning a star wheel, reached either through a window in the backplate or in the drum. Remember too that drum brakes on American cars may well be self-adjusting both on the front and at the rear, so before the star wheel can be turned, a screwdriver must be inserted to push the adjusting lever out of engagement. Typical operations are shown in Figs. 4:1 and 4:2. In the case of rear brakes, it may also be necessary to relieve all the tension from the handbrake, if the cables are tight; usually a matter of loosening off the adjustment nut at the cable equalizer.

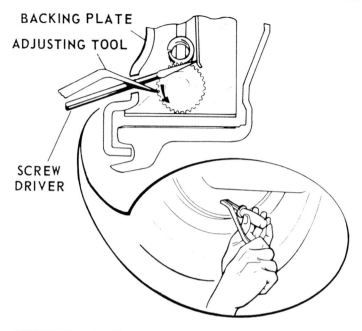

BACKING PLATE

ADJUSTING TOOL

SCREW
DRIVER

Fig. 4:1. Typical method of
backing off brake adjustment
through an aperture in the
backing plate using a
screwdriver

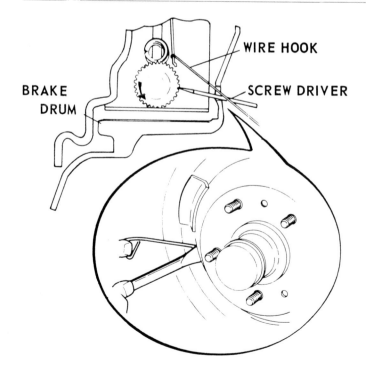

WIRE HOOK

BRAKE
DRUM

SCREW DRIVER

Fig. 4:2. Typical method of
backing off brake adjustment
through a hole in the drum

The drums come off in the same way as on British and European cars, although it is likely that the front ones, particularly, will involve pulling the split pin, taking off the castellated axle nut to free the combined drum and hub, and that may well mean that the bearing assembly comes apart. It is also possible that a rear drum might have a locating tag. Note it if it has, but mark the drum and one of the studs, if not, to ensure the drum goes back into the same location from which it was removed.

Wheel cylinder clamps are recommended in American maintenance manuals, but, provided the brake pedal is not depressed while working, a couple of stout elastic bands wrapped around should prevent air getting in.

Return springs on American drum brakes tend to be a bit more complicated than those on Europe's smaller cars (Fig. 4:3 is typical Bendix), so it pays at this stage to note carefully the colour of each spring and exactly where it is located. Draw pictures, write notes, or leave one side assembled as a reference, but make sure you know how to put everything back afterwards.

Start dismantling by unhooking the shoe return springs (5). The Americans use a special tool but as usual it can be done without. Follow these with the hold down springs (8). These may well be the same as those used in Europe and can be tackled with the special tool if it's available or with a pair of pliers if it isn't.

Spread the shoes slightly and extract the handbrake strut (13) and then disconnect the cable from the handbrake lever (12).

The shoes are removed by first lifting the anchor plate (4) off its pin, freeing the shoes from the wheel cylinder and removing the shoes together with the adjustment components as an assembly, still connected at the bottom by adjusting screw and spring.

On the bench, these last two can be removed from the shoes, followed by the parts from the self-adjusters mechanism (if they're fitted), and that's the adjusting lever (15), cable (16), and cable guide (17).

Cleaning and inspection come next. If the brake shoes are unevenly worn, it could indicate drum defects—either tapered or running out of true—and there is more about

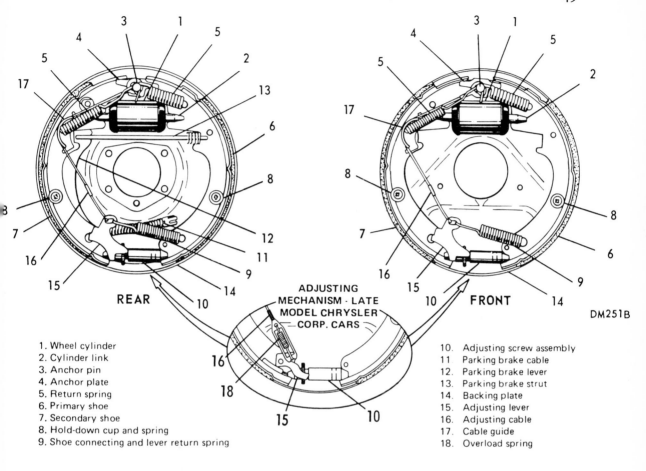

1. Wheel cylinder
2. Cylinder link
3. Anchor pin
4. Anchor plate
5. Return spring
6. Primary shoe
7. Secondary shoe
8. Hold-down cup and spring
9. Shoe connecting and lever return spring

10. Adjusting screw assembly
11. Parking brake cable
12. Parking brake lever
13. Parking brake strut
14. Backing plate
15. Adjusting lever
16. Adjusting cable
17. Cable guide
18. Overload spring

that in chapter seven. If the friction materials are oil soaked, check the axle oil seal and the wheel cylinder (check that anyway by lifting the edge of the dust cover and looking for leaks). If it is leaking, or if one piston (or both) are seized, which could show up as wear or lack of it on the linings, or if the rubber is cracked or damaged, you'll either need to rebuild it with a kit or (more likely) fit a replacement cylinder.

Clean up the backplate by scraping and brushing loose dirt off with a wire brush and preferably collecting it in a vacuum cleaner. Make sure it doesn't fly about; breathing asbestos dust can damage your lungs. Use a cleaning fluid from one of the brake companies to make sure both the

Fig. 4:3. Bendix self-adjusting brake assemblies used on typical large and intermediate sized American cars

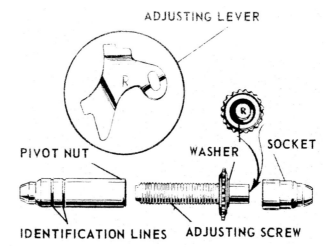

ADJUSTING LEVER

PIVOT NUT   WASHER   SOCKET

IDENTIFICATION LINES   ADJUSTING SCREW

Fig. 4:4. The main requirements on this adjusting screw are to ensure it is clean, lubricated and free moving

backplate and other parts that are to be reused are free of dirt. If the raised platforms on the backplate against which the shoes move are corroded, use emery cloth to clean them off and smear thinly with brake lubricant.

Take the adjusting screw apart (Fig. 4:4) and clean it up in meths or cleaning fluid. Run the nut from end to end of the thread to make certain it's free and ensure none of the teeth on the adjuster wheel are damaged. Use brake lubricant sparingly on the thread and reassemble, ensuring no grease gets on the adjuster teeth. If an 'anti-noise' washer was fitted originally, make sure it is replaced. Note that the nut on one of the adjusting screw assemblies has a left-hand thread. It is important therefore that they are not transposed across the axle.

Sort out which is the primary and secondary shoe of your replacement pair (they are different) and then use emery cloth to remove any burrs from the shoes where they contact the raised platforms on the backplate. Reassembly is now the reverse of dismantling.

On rear brakes, refit the handbrake lever (12) to the secondary shoe using the spring washer and retaining clip. Fit both shoes to the backplate by means of the hold-down springs (8). Refit all the self-adjusting mechanism and the shoe retracting springs, and when fitting the adjusting screw at the bottom, ensure the toothed wheel is at the secondary shoe end.

When it's all assembled, check the action of the automatic adjusting mechanism by pulling the cable between the cable guide and anchor pin far enough to lift the lever past one tooth of the adjusting wheel. Check that it snaps firmly into place behind the next tooth, and then release the cable, when the lever should return to its original position, turning the wheel one tooth as it does so.

If it doesn't work properly, it is possible that a new cable is needed or that there is damage to the cable guide or the lever.

Adjust the brake so it is just possible to refit the drum and then make final adjustments through the window either in the drum or backplate.

As with any other brakes, shoe replacement should always be done 'across the axle' otherwise uneven braking could result.

Disc brakes

The actual disc brake chosen here, and shown in Fig. 4:5, is a Delco Moraine four-piston design. Used on, for instance, Buick, Chevrolet, Oldsmobile and Pontiac, it is similar in principle (if not in detail) to the Bendix, Budd and Kelsey Hayes types used elsewhere.

For a pad change/inspection and overhaul, the pre-

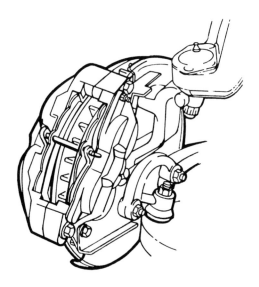

Fig. 4:5. This Delco Moraine four-piston disc brake has been used by Buick, Chevrolet, Oldsmobile and Pontiac and is similar in principle to many others

liminary steps are the usual ones of jacking the car up, supporting the relevant end on axle stands and removing the road wheels.

If you're going to fit new pads—and it's generally a good idea—start by removing two thirds of the fluid from the master cylinder. Probably the easiest way is to open the bleed nipple on the brake you're working on, depress the brake pedal and force it out into a jar. The rather more messy alternative is to take the master cylinder cap off and wrap old rags around the outside, the idea being that these will soak up fluid as it overflows when the pistons in the caliper are pushed back.

Push the inboard shoe away from the disc (you can do it with your fingers) and if you can beg, borrow or steal some, slip retaining clips (J-22674) over the caliper half, with the flat side between the back of the pad and the caliper to protect the pistons (Fig. 4:6). Pull out the split pin from the inboard end of the shoe retaining pin and throw it away and then slide out the retaining pin.

Take out the old shoes. On some brakes they lift straight out, but others have to be rotated one way or another to get one end out and then can be lifted clear (Fig. 4:7). Go through this part of the operation carefully, noting, for instance, if there are any shims and how they're fitted,

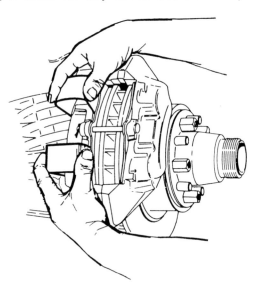

Fig. 4:6. These retaining clips are recommended while pads (the Americans call them shoes) are removed

QUICK CHANGE SHOE
1967–1968   2000 SERIES DISC BRAKE

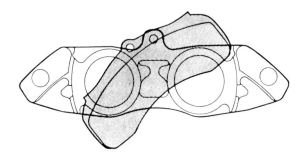

Fig. 4:7. On some cars equipped with Delco Moraine discs, the pads have to be rotated in order to remove them. This does not apply to the Corvette

because they've got to go back the same way. Another point is that inboard and outboard shoes are different. There should be an arrow on the back of each new shoe and this must point in the direction of forward disc rotation if they are to be correctly fitted.

While the shoes are out, check for piston seal leaks, cracked caliper castings and any damage to the dust covers. Clean up the whole area if all is well; if it isn't, you might have to fit a new caliper, or at least recondition the old one.

Check the disc surface for any signs of rust building up and if it is corroded, spin the disc against a hand-held screwdriver to chase it off, and finally clean up with emery cloth.

With the new pads fitted, refit the retaining pin from the outside of the caliper. Lock it with a new stainless steel split pin on the inside of the caliper. This should be supplied with the new pad kit.

Finally, refill the master cylinder reservoir with the correct fluid and when both sets of pads have been renewed and both calipers checked, work the footbrake pedal until it becomes hard; make sure you do that *before* driving the car.

If a caliper hydraulic overhaul is required, the general information in chapter 1 may be used, but one interesting difference with this unit is that, unlike the Girling units, it may be split into its two halves and you can even replace only one of them, should damage to the cylinders make it necessary.

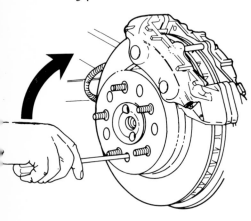

Fig. 4:8. On this rather different drum parking brake fitted to a rear disc brake on the Corvette, adjustment is carried out as shown

Rear disc brake

It is not proposed to go into a great deal of detail, but the brake shown on Fig. 4:8 and fitted to the Corvette is a departure from the drum brakes fitted mostly to the rear of American cars. The main rear foot brake is a four-piston caliper operating on a disc, but the handbrake (parking brake in America) is a small drum system incorporated in it.

In order to get at the brake shoes, the whole assembly has to come off. This involves first taking out the disc pads and then removing the caliper. Attention is then turned to the front face where the road wheel is normally fitted. Mark one of the wheel mounting studs and the appropriate hole on the face to facilitate identical assembly. It is then a case of drilling out the five rivets and removing the disc assembly (Fig. 4:9). There is no need to obtain new rivets incidentally; they were originally used for manufacturing purposes at the factory and on reassembly the whole thing will be held by bolting the wheel to its studs.

Servicing the parking brake shoes follows the same procedures outlined earlier in the chapter and in the drum brake sections of the preceding three chapters. All the component parts are similar to those mentioned earlier, the adjustment mechanism is the same and the main essential is to note the position of everything very carefully before dismantling, so that it all goes back together properly afterwards. Don't forget, after cleaning the backplate, to lubricate the raised platforms (six of them) against which the shoes slide, and to clean and lubricate the adjuster. It's turned manually via a hole between two of the wheel mounting studs.

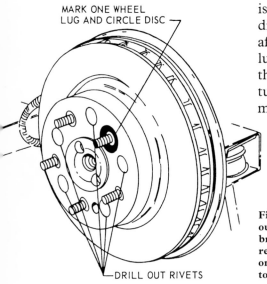

MARK ONE WHEEL
LUG AND CIRCLE DISC

DRILL OUT RIVETS

Fig. 4:9. Rivets have to be drilled out to gain access to the parking brake, but there's no need to renew them because the wheel on its studs holds everything together

Chapter 5 | Hydraulics overhaul

Note: The use of a service kit can never be as safe or satisfactory as replacing an old unit with a new guaranteed one. If in doubt as to the suitability of resealing, get expert advice.

When hydraulic components fail, there are usually two alternatives—either to replace the part or overhaul it using a maker's kit. The difference in cost between the two is so great that it is often found expedient to try the effect of rebuilding with new seals first and then, if that doesn't work, replace the whole thing. The exception to this is possibly drum brake wheel cylinders, where the fairly low replacement cost does relieve some of the burden. The biggest job is likely to involve the master cylinder, however, so they will be dealt with first.

Master cylinders
Generally, location is fairly standardized and most manufacturers tend to mount it on the bulkhead under the bonnet in line with the top pivot of the brake pedal. This obviously gives direct action by means of a short operating rod, but it doesn't make the job of changing the unit too convenient.

Start by draining the brake fluid from the master cylinder reservoir. The simplest way of doing this it to connect a bleed tube to the nipple on one of the front brakes, undo the nipple and pump the fluid out into a jar. When the level has dropped far enough, close the bleed nipple again.

Inside the car is the awkward bit where you have to lay flat on your back under the steering wheel in order to pull out the split pin and withdraw the clevis pin. It's even more awkward when refitting.

Back under the bonnet, it is simply a matter of disconnecting two hydraulic pipes (more if it's dual circuit) and the mounting bolts on the master cylinder unit itself.

Take care when lifting it out not to drip brake fluid over the paintwork—it's a good paint stripper!

There are exceptions to this procedure, some easier and others more difficult. The easier way is when the master cylinder is bolted to a direct servo, when only the under the bonnet part of the operation and the fluid draining exercise have to be carried out.

It's more difficult in some cases where the master cylinder is mounted elsewhere, like the well known case of the Morris Minor where it's under the floor, or the Hillman Imp where it's under the front-mounted petrol tank. Access in both these cases is a lot more complicated.

When access is this difficult, it does provide another reason for not rebuilding the unit and hoping for the best, but instead fitting a replacement unit right from the beginning. In any case where a replacement unit is being installed, the operation is simply the reversal of the removal procedure.

The alternative, of course, is to rebuild the unit with an overhaul kit, but the success of this does depend on the condition of the cylinder itself and on doing the work carefully and in conditions of scrupulous cleanliness.

Start by clearing an area of the bench and covering it with paper; something plain is better than newspaper and the back of an old roll of wallpaper would do quite well. Wash the outside of the master cylinder in cleaning fluid before starting dismantling, making sure you use the brake company's own proprietary cleaning fluid for the job. This will not affect the rubber parts.

Probably the most typical Girling master cylinder of the postwar years is the C.V. (centre valve) type shown in Fig. 5:1. Start dismantling by pulling back the rubber dust cover and removing the circlip with a pair of long-nosed pliers. This releases the pushrod which can be pulled clear together with the dust cover. Then by shaking or tapping the cylinder on the bench, the plunger assembly can be extracted.

Now is a good time to compare the seals on the old assembly with those supplied in the overhaul kit. If the old ones appear oversize, contamination is the most likely cause, and if those in the master cylinder are ruined, so

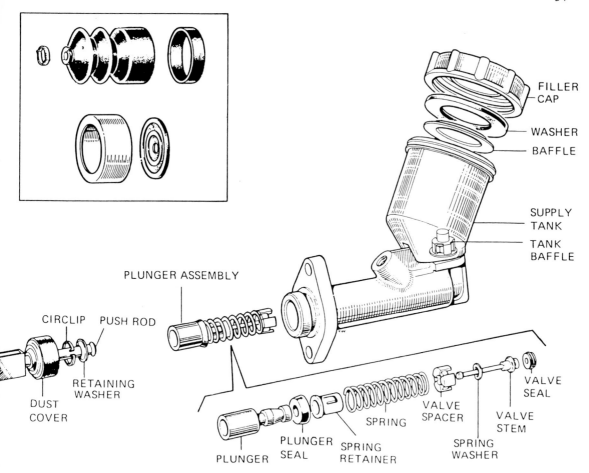

FILLER CAP

WASHER

BAFFLE

SUPPLY TANK

TANK BAFFLE

PLUNGER ASSEMBLY

CIRCLIP PUSH ROD

RETAINING WASHER

DUST COVER

PLUNGER

PLUNGER SEAL

SPRING RETAINER

SPRING

VALVE SPACER

SPRING WASHER

VALVE STEM

VALVE SEAL

probably are all the rest in the system. This would mean not only changing every seal, but also the flexible hoses, and careful flushing through with cleaning fluid.

To dismantle the plunger assembly, lift the bent-in leaf of the spring retainer (Fig. 5:2) and ease the spring assembly from the plunger. Then compress the spring and free the valve stem from the keyhole of the spring retainer. The spring, valve spacer and spring washer can then be separated from the valve stem and the valve seal from the valve head.

Now look at the plunger. There are two types, as shown in Fig. 5:2. If you have type A, the seal which isn't deeply

Fig. 5:1. The Girling C.V. (centre valve) master cylinder was very frequently used in the postwar years.

Adjustable length pushrods, where fitted, should be set to give $\frac{1}{32}$ in. free play minimum at the master cylinder (approximately $\frac{1}{4}$–$\frac{5}{16}$ in. at the brake pedal)

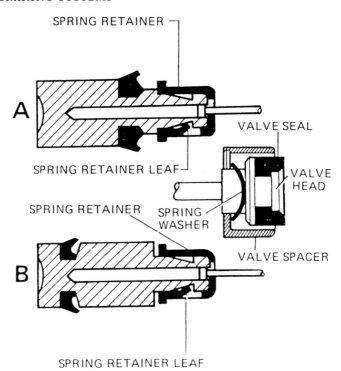

SPRING RETAINER

A

VALVE SEAL

VALVE HEAD

SPRING RETAINER LEAF

SPRING RETAINER

SPRING WASHER

B

VALVE SPACER

SPRING RETAINER LEAF

Above **Fig. 5:2. Two types of plunger assembly shown in detail**

recessed can simply be pulled off. If, however, the seal is situated at the pushrod end of the plunger, as at B, you'll need to take special care. Find a small screwdriver and round off all the corners and edges at the tip. The idea is to prevent it from damaging the plunger; if it does, a new master cylinder must be fitted.

The method of getting the seal off can be seen in Fig. 5:3. The seal is squeezed until the smooth end of the screwdriver can be slipped underneath and carefully levered out of its groove and pulled off the plunger using the fingers as much as possible.

You can see from the parts contained in the overhaul kit what needs replacing and what has to be cleaned up and refitted. For these latter use fresh brake fluid, dry them and lay them on a clean surface.

Opposite **Fig. 5:4. The tandem master cylinder is similar in many ways to the normal single type, but with the addition of a primary plunger assembly and a tipping valve assembly**

The critical bits to look at are the plunger and the cylinder bore. There must be no visible score marks, wear ridges or corrosion. Check the bore particularly; it should be completely smooth to the touch. If any damage is discovered, a new unit must be fitted.

Start reassembly by fitting the new plunger seal. You can see from Fig. 5:2 which way round it goes. Use clean brake fluid as a lubricant and do the job on a clean surface and with clean hands.

Fit the valve seal with the smallest diameter end going onto the valve head first. Follow this with the spring washer, ensuring it curves away from the valve stem shoulder (Fig. 5:2), then push on the valve spacer, legs first and the spring. Follow this with the spring retainer, compressing the spring until the end of the valve stem can be passed through the centre hole and locked. Then attach this sub-assembly to the plunger and lever down the leaf of the spring retainer to hold it in place (Fig. 5:2).

Use plenty of clean brake fluid to lubricate the assembly and the inside of the cylinder and then insert it, easing the seal at the entrance. Position the pushrod and retaining

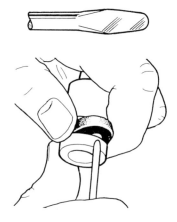

Above **Fig. 5:3. This is the method used to remove the seal without damage to the plunger, using a small screwdriver with the corners of the bit rounded off**

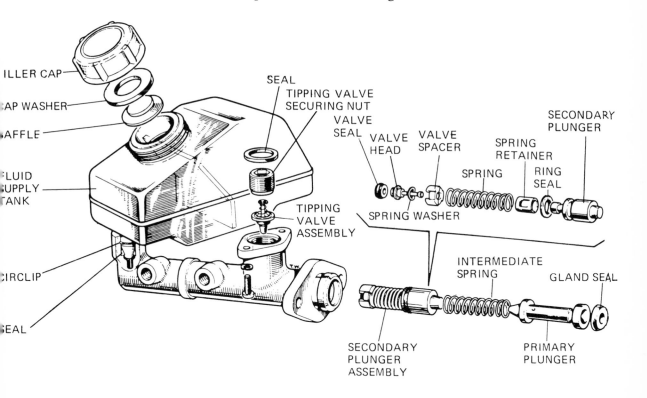

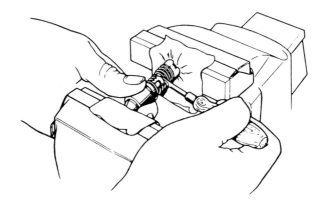

Fig. 5:5. When reassembling, with the spring compressed, the spring retainer must be pressed right back against the secondary plunger with a small screwdriver

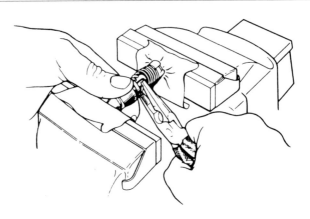

Fig. 5:6. Use needle nose pliers to ensure the spring retainer leaf locates squarely behind the plunger head

washer and then use the thin pliers to engage the circlip. Check everything is locked in place. Use rubber grease to lubricate the pushrod and fill the dust cover. Finally refix the dust cover, depending on what type is used.

Tandem master cylinders started to appear in the mid-1960s and might be a little modern for the attentions of a restorer, but they were beginning to be used 20 years ago, so it might be as well to know what to expect. In the case of Girling units, they are not so very different from the single circuit types. As can be seen in Fig. 5:4, the secondary plunger assembly closely resembles the one in the single master cylinder; the main additions are the simple primary plunger and the tipping valve assembly.

The overhaul process using a Girling kit is very similar too and the only additional complication perhaps is that the spring has to be compressed when the leaf of the spring retainer is pressed down behind the head of the plunger. A bench vice is best used for this, with two pieces of paper to protect the hydraulic parts against dirt. Two points must be made; first, with the spring compressed the spring retainer must be pressed right back against the secondary plunger with a small screwdriver (Fig. 5:5), then use needle nose pliers to ensure that the spring retainer leaf is squarely located behind the plunger head (Fig. 5:6). Note that the tipping valve securing nut should be tightened to a torque setting of 35–40 lb ft. All the rest of the overhaul procedure follows closely along the lines of the single unit.

Lockheed master cylinders come in several shapes, but the internal mechanism of all of them is much the same. It is simpler than the Girling type in that taking off one circlip releases all the parts and there are no internal assemblies to build up before reassembling.

Just like the Girling, overhaul kits are available and cross checking with the kit will underline which parts need replacing and which are re-used. All the same strictures about cleanliness apply. Only brake company approved cleaning fluids can be employed (never petrol or paraffin), only clean brake fluid is permissible as an assembly lubricant inside the unit and only proper brake grease is used on the operating rod and dust cover—'Rubberlube' in this case.

Most other master cylinders for single circuit systems are much the same, and there is no need to know how they work in order to be able to overhaul them. Usually there are instructions of some sort with the overhaul kit and if a diagram can be obtained from the relevant workshop manual, applying all the foregoing general advice should enable the work to be done without difficulty. If you don't have a diagram to detail the order of assembly, it's mainly a matter of dismantling in strict sequence and noting every item.

Wheel cylinders

Compared with master cylinders, wheel cylinders are simple. They come in different shapes and sizes to suit the backplates of a wide variety of vehicles, but in basic terms

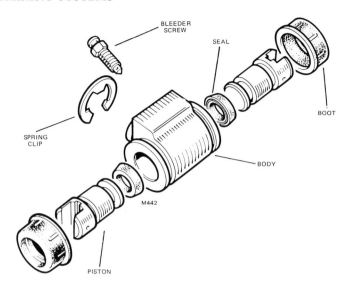

BLEEDER
SCREW

SEAL

BOOT

SPRING
CLIP

BODY

M442

PISTON

Fig. 5:7. Shown here is what is probably the simplest type of wheel cylinder fixing, but although the spring clip is the least complicated it is not the easiest to handle. Also from this illustration it is possible to see typical construction and internal wheel cylinder detail

they are either single acting or dual acting, and the main complications come more from the way they are fitted and perhaps in reassembling the automatic adjustment mechanisms which in some cases are mounted on them.

Dealing with changing a wheel cylinder for a new replacement first, access has already been dealt with in the drum brakes chapter, and basically it means removing the drum and the brake shoes. Before tackling the wheel cylinder itself, it's a good idea to minimize the loss of fluid and the ingress of air by clamping off the adjacent brake hose. Use a proper brake clamp for this if you can, which will not harm the flexible brake hose, but if you don't have one, a pair of $\frac{5}{16}$ in. dia. rods clamped with a locking wrench will do the trick.

Take off the cylinder next by first disconnecting the hydraulic pipe and, if you can, plug both the entry into the cylinder and the end of the pipe. Take out the bleedscrew and also plug that.

The mounting of the wheel cylinder may have been achieved in a number of different ways. The simplest is one or more set screws, or studs with springs washers and nuts. Another system is the Lockheed one of a single flat spring clip (Fig. 5:7) which fits firmly into a groove on part of the cylinder that locates through a hole in the backplate. This system has its problems first in getting the old spring clip out, particularly when it is rusty, and secondly in springing in the new one. Lockheed always recommend fitting a new clip, and the old one is often so twisted and battered by the time it's extracted that this is the only way it could be done.

Other spring clip methods are those used by Girling. One is a pair of opposed flat U-clips, an inner and outer, which are simply driven off in opposite directions (Fig. 5:8), while another is slightly more complicated, involving an oval-shaped spring clip and two Y-shaped plates. The sequence of removal and refitting is shown in Fig. 5:9.

If the new cylinder is not fitted with a bleed nipple, take the one out of the old cylinder, or alternatively get a new one. Leave the hydraulic connection hole plugged, until you are ready to refit the pipe. Clean up the backplate on both sides before installing and coat with grease where a

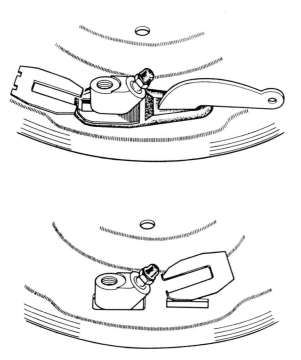

Fig. 5:8. Another common type of wheel cylinder mounting method. This one is used by Girling and consists of a pair of opposing U-clips

a

When fitted correctly, the spring clips should look like this.

c

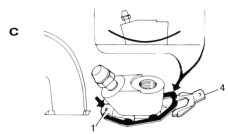

Press down the plate (1) and ease out plate (4).

b

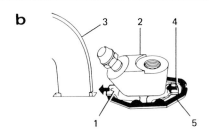

To remove the cylinder from the backplate, slide the 'Y' shaped plate (1) as far as possible away from the turret of the cylinder (2) and towards the lever (3). Slide the second 'Y' shaped plate (4) inwards towards the turret (2) so that the leg can be eased under the spring clip (5).

d

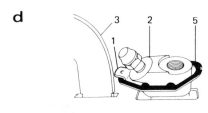

Lift the plate (1) and the clip (5) off the turret (2). Work out the lever (3) then the wheel cylinder can be removed. To fit a new cylinder, clean the backplate on both sides where the cylinder slides and apply Girling Brake Grease, then reverse the removal procedure. When fitted, it is important not to slide the cylinder more than is absolutely necessary until the brake shoes and drum are fitted as the plates may become disengaged.

Fig. 5:9. Another Girling system utilizes an oval shaped spring clip and two Y-shaped plates. Note the sequence of fitting and removal

sliding type cylinder is used. The remainder of the installation work is the reversal of removal.

Overhauling a wheel cylinder follows exactly the same techniques as with the master cylinder. Dismantle it, thoroughly clean all the parts in clean brake fluid, and check both cylinder walls and the piston surface for signs of corrosion, wear or scoring. Provided there is nothing wrong, fit the new seal from the kit to the piston, lubricate the bore with brake fluid, and the piston, and insert it. Fit the new dust cover and retainer. Finally screw in the cleaned-up bleed screw.

Caliper hydraulics

Problems with the hydraulic side of disc calipers usually first come to light when carrying out a pad change. It may be dust seals in a bad state or it may be a fluid leak. It could be that the pads are unevenly worn, indicating a seized piston, or perhaps binding in the moving part of a sliding caliper. Whatever the cause, if it involves a hydraulic overhaul, in many cases it means removing the caliper.

With the Girling caliper, before doing this, remove the brake pads as described in the earlier chapter. Fit a block of wood in place of one of them and gently depress the brake pedal. This will bring the free piston out of its housing. When it's just far enough to be gripped, block it from moving further, remove the packing from the other side and work the pedal again to partially eject the second piston.

Drain the hydraulic fluid by opening the bleed nipple a half turn and pumping from the brake pedal. Disconnect the brake from the hydraulic circuit by the most convenient means. If a brake pipe connects directly to the caliper, undo this union. If the connection is by a flexible hose, make the disconnection where this joins the circuit at a bracket union. Undo the bolts which hold the caliper to the stub axle and remove it.

On the bench, check out which type of dust covers are fitted, and you can either remove them first and then the pistons, or the other way round, depending on type (see Fig. 5:10). Provided the pistons have been ejected far enough before removing the caliper, it should be relatively easy to pull them right out; otherwise the recommended method is to use compressed air, which not everyone has, of course.

Once the pistons are removed, the sealing rings can be picked out of their grooves in the cylinder walls. Take care not to damage either cylinder or groove.

Clean up pistons and bores using Girling cleaning fluid or clean brake fluid and check for damage, abrasion, scuffing or corrosion. If a piston is suspect, it may be replaced, but if there is damage to the cylinder walls, a new complete caliper must be obtained.

Lubricate the inside of the cylinders with clean brake fluid, then do the same for the new sealing rings and fit them carefully. Depending on which type of dust covers you have, these will either be fitted first, followed by the pistons, or the other way round. With A and B, the piston

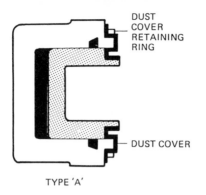

DUST
COVER
RETAINING
RING

DUST COVER

TYPE 'A'

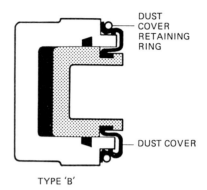

DUST
COVER
RETAINING
RING

DUST COVER

TYPE 'B'

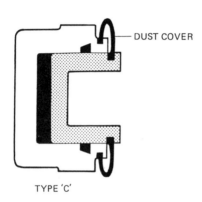

DUST COVER

DUST COVER

TYPE 'C'

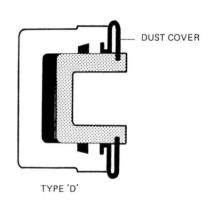

DUST COVER

DUST COVER

TYPE 'D'

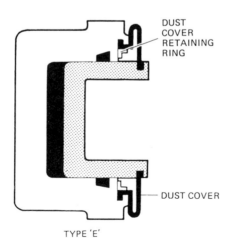

DUST
COVER
RETAINING
RING

DUST COVER

TYPE 'E'

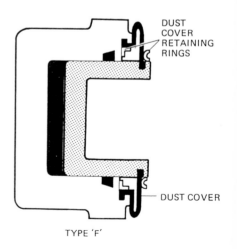

DUST
COVER
RETAINING
RINGS

DUST COVER

TYPE 'F'

goes in first and then the dust cover and retaining ring are fitted. With C, the dust cover goes on the piston first, then the piston is inserted and the other lip of the dust cover located in its groove in the body. For type D the cover is located in the body and the piston inserted through it, finally being located in the piston. Type E is fitted in a similar sequence to D, care being taken to locate the retaining ring using two screwdrivers to avoid distortion. The piston goes in next and finally the dust cover located in the piston groove. F is similar but with an extra retaining ring.

When refitting the caliper, make sure any shims originally included are refitted exactly as they were. Finally bleed the system as described in chapter eleven.

The procedure with the Girling 'A' type caliper, fitted to

Above Fig. 5:11. This is a useful trick. Trapping a sheet of polythene under the master cylinder cap limits the amount of escaping fluid when working on the hydraulics anywhere in the braking circuit

Opposite Fig. 5:10. The range of dust seal types used on Girling disc brake calipers

Fig. 5:12. Dismantling a master cylinder begins by pulling out the operating rod and releasing the circlip in the end of the body

Maxi and Allegro, is a bit different. Remove the caliper from the car and clamp it in a vice. Press the top piston right back into its cylinder and then push the cylinder body downwards to separate it from the yoke.

Retaining rings and dust covers come off next, but then, unless you have compressed air in your workshop, it'll mean a trip to the nearest friendly garage to borrow their air line to eject the top piston before following this with the lower one. Back on the bench, the sealing rings can be extracted from their grooves in the cylinder walls.

Checking the condition of cylinders and pistons follows the sequence already described, and so also does the actual hydraulic side of rebuilding. There are, however, some differences in reassembling the caliper due to variations of

Fig. 5:13. A Girling master cylinder is dismantled. Note the clean bench and utensils, and even the hands are clean

design, but these will have been noted when dismantling.

With the Lockheed horizontally-opposed piston design, the makers recommend a slightly different technique. Their instructions are to take the caliper off and support it without disconnecting it hydraulically. With the pads removed, clamp one piston and gently use the brake pedal to eject the other. A blunt screwdriver levers out the wiper seal retainer from the cylinder rim, followed by the wiper seal and the pressure seal further in. Obviously great care must be taken not to damage the cylinder walls. Brake fluid or ethyl alcohol must be used to clean everything and checking follows along the same lines as already described.

The inner pressure seal is fitted first so that it feels proud of its groove on the side furthest from the mouth of the

Fig. 5:14. This is one of the Girling systems for securing wheel cylinders, used on a great many cars in the postwar period

bore. The piston goes in next, but leaving $\frac{5}{16}$ in. protruding and then the new wiper seal is fitted into its retainer and the protruding piston used as a guide for fitting. The bleed screw can be opened slightly to facilitate pushing the piston in, and the clamp mentioned earlier used to locate the seal and retainer squarely and without damage, and to push home the piston. The entire procedure is then repeated on the other half of the unit.

With the Lockheed single cylinder (swinging) caliper, the recommended hydraulic overhaul procedure follows very much along the lines of the previous Lockheed caliper, but there are, of course, some differences dictated by the different design. The following is the procedure in brief.

With the caliper removed, supported, but not discon-

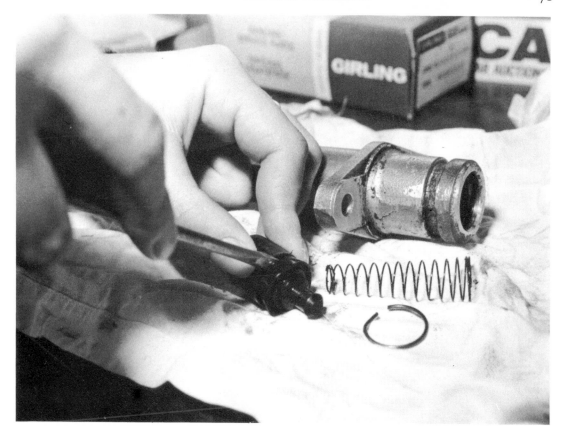

nected hydraulically, the piston is almost ejected by pedal action. The unit is then removed, and by taking off the spring clip(s) the cylinder assembly is separated from the frame. Everything is cleaned up and the piston taken out. From this point on the work is very similar to the instructions for the previous caliper, except, of course, there is only one piston.

Disc brakes of different makes are unlikely to show much variation in design to those already covered in some detail.

It should also be possible by this token to dismantle them and carry out a hydraulics check in exactly the same way. Ensure before you start, however, that the relevant overhaul kits, either seals, or pistons and seals, are all obtainable.

Fig. 5:15. Seals can be obstinate but must be removed with great care to avoid damaging pistons, etc. Here a blunted screwdriver is used as a lever

Chapter 6 | Servos—service and overhaul

Basically there are two types of servo unit—hydraulic and mechanical. They both do the same job of using vacuum power from the engine to boost hydraulic pressure in the braking system, but the main difference between them is that the hydraulic servo is fitted between the master cylinder and the rest of the circuit, while the mechanical servo is mounted between brake pedal and master cylinder.

This latter system tends to restrict the number of places it can be placed and usually it is fitted on the bulkhead in exactly the same place as the master cylinder is normally mounted. Occasionally, when a model is converted from left-hand drive, for instance, the servo is left where it was originally installed on the left-hand side and connected mechanically across to the brake pedal. Sometimes, if there is no room in the orthodox position for a mechanical servo, it is possible that a hydraulic servo, which can be fitted literally anywhere, will be installed instead.

Whereas home overhaul of much of the braking system presents little problem, the servo is not quite so simple. A certain amount of very basic maintenance is possible on all of them, but when it comes to anything more comprehensive, often the best recommendation is to buy a complete replacement unit.

The servo does tend to go on working for a very long time without giving any trouble; after a very high mileage, however, it may not be a case of just a simple fault, but instead a new unit might well be a better solution. 'Crunch mileage,' according to one manufacturer (Girling), is 40,000 miles, 64,000 km or three years, both for the hydraulic servo, for which a major service kit is available, and for the mechanical unit, which they definitely recommend should be changed, and for which no major service kit is supplied, only some smaller ones.

Hydraulic servos

There are some basic tests which can be done to help decide whether a hydraulic servo actually needs an overhaul or not; most people tend to let them work on and only do something about them when a fault develops. To see whether it is operating, have a look at all the connections first, particularly the vacuum manifold pipe. Start the engine and get an assistant to push the brake pedal. You should be able to hear air going through the inlet and, by placing a hand on the vacuum chamber of the unit, to feel it operate.

With the engine still running, apply the brake with a steady pressure for about half a minute. The pedal should stay firm and not sink (even very gradually) to the floor.

With the engine running and a vacuum built up, stop the engine, leave for two minutes, then try the brakes. At least two applications should remain vacuum assisted.

Exhaust the vacuum within the servo, then depress the brake pedal and keep it pushed down while the engine is started. The pedal should drop slightly and then remain firm again.

If you can't hear air entering or feel the operation, the unit probably isn't working at all. Try changing the non-return valve, for which there is a kit, and fitting a new vacuum hose. This can be done without removing the unit. If the pedal tends to sink, leaks or scored bores internally are likely and the unit will have to be changed.

About the only other service job that can be done is to replace the air filter, and this is supposed to be renewed every 12,000 miles or so, or when new brake shoes and pads are fitted. Depending on make, several different methods have been used to secure the filter, but it is always just a simple service job.

If you have something a lot more ambitious in mind, like a complete overhaul, check first with your local supplier to find out if you can get the relevant repair kit. Do nothing until you've got it, and don't break the seals on the kit to see what's inside. Leave this until you've removed the unit and dismantled it. If you need a new unit after all, this way you can get a refund on the overhaul kit.

Overhaul

The same care must be taken when working with a servo as with the rest of a hydraulic system. Only proprietary brake manufacturer's own cleaning fluid can be used, and immense care must be exercised to ensure no dirt contaminates the internal workings of the unit – clean hands, clean bench, clean tools, clean everything.

We have chosen here to describe in detail one of the most commonly used servos – the Girling Mk. 2B unit (Fig. 6:1). Fitted as original equipment to many cars, it was also marketed in accessory form as the Girling Powerstop. It is essentially the same as the earlier Girling model and 2A and operates on much the same principles as many other servos, so much the same procedures can be used to overhaul units of other makes.

Clean the outside of the servo unit before removing it from the car and note the hydraulic connections as you undo the unions; then undo the mounting bolts and lift the servo out.

A slight complication arises right at the beginning of dismantling. It's not a good idea to clamp the unit in the vice by one of its mounting lugs because damage would mean fitting a new unit. You need a support plate similar to that shown in Fig. 6:2. Use the actual mounting lugs to mark the centres for drilling 9 mm ($\frac{3}{8}$ in.) holes, and ensure the other dimensions marked are adhered to, particularly the thickness of 6 mm ($\frac{1}{4}$ in.)

Bolt the servo body to the support plate and mount it in the vice so that the little plate welded to the clamping ring on the unit is uppermost (Fig. 6:3). Scribe a line across the two halves of the unit close to the plate so they can be reassembled in the same position.

Another tricky operation here, and hopefully it is possible to get the car relatively close to the bench so a vacuum hose can be connected between the non-return valve on the bench-mounted unit and the engine's inlet manifold stub. Starting the engine will then draw the two

Fig. 6:1. The Girling Mk 2B hydraulic servo unit exploded into all its component parts

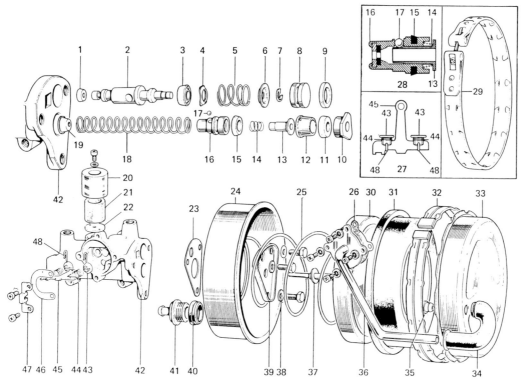

KEY

1. CONTROL SEAL (SECONDARY)
2. CONTROL PISTON
3. CONTROL SEAL (PRIMARY)
4. SPRING ABUTMENT
5. SPRING
6. RETAINER
7. CIRCLIP
8. PLUG
9. PLUG SEAL
10. BUSH
11. PISTON ROD GLAND SEAL
12. SPACER
13. SLEEVE
14. PISTON SLEEVE SPRING
15. OUTPUT PISTON SEAL
16. ANTI-KNOCK OUTPUT PISTON

17. BALL
18. OUTPUT PISTON SPRING
19. PISTON STOP
20. COVER
21. FILTER ELEMENT
22. FILTER BASE WASHER
23. BODY GASKET
24. REAR SHELL
25. DIAPHRAGM RETURN SPRING
26. COVER GASKET (VALVE CHEST)
27. 'T' LEVER VALVE ASSEMBLY
28. ANTI-KNOCK OUTPUT PISTON ASSEMBLY
29. SERVICE CLAMPING RING
30. DIAPHRAGM PLATE & PISTON ROD ASSEMBLY

31. DIAPHRAGM
32. CLAMPING RING
33. FRONT SHELL
34. RUBBER SLEEVE
35. PLUG
36. VACUUM PIPE
37. ABUTMENT WASHER
38. (COPPER) WASHER
39. CLAMPING PLATE
40. GROMMET
41. NON-RETURN VALVE/ADAPTOR
42. HYDRAULIC BODY
43. VALVE
44. VALVE SPRING
45. 'T' LEVER
46. SPRING PLATE
47. LEVER GUIDE
48. SPRING CLIP

halves of the shell together.

Cover or blank off the hydraulic ports so prevent filings from entering (plastic plugs or a sheet of polythene film will do), and hacksaw through the centre of the clamping ring plate, taking great care not to damage the two shells. Use a screwdriver to lever the clamping ring off.

Lean heavily against the front shell of the unit and at the same time undo the setscrews to release the valve chest cover. Air can then enter the unit and break the vacuum lock, and you'll need to maintain pressure on the cover or the two halves will fly apart under pressure from the diaphragm return spring.

Remove the spring and cover and turn off the car engine.

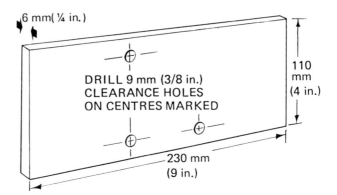

Fig. 6:2. Dimensions for making a support plate—an essential for satisfactory dismantling

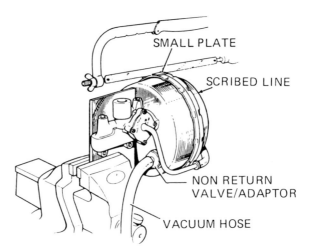

Fig. 6:3. Hacksawing through the plate on the clamping ring so the two halves of the shell can be parted

Remove the vacuum hose, pull out the vacuum pipe, and use a Phillips screwdriver to lever the support plug out of the vacuum sleeve and finally lever the sleeve out of the front shell with an ordinary flat screwdriver (Fig. 6:4).

Using the support plate again, put the unit back in the vice and undo the three bolts holding the rear shell to the body (Fig. 6:5). Lift off the triangular clamping plate carefully as the output spring underneath may eject all the various parts from the bore. The trick is to hold the piston down with a finger and release the pressure slowly to control the release of the anti-knock piston and the seals. If they aren't ejected spontaneously, withdraw the bush and hook out the piston rod gland seal and this will enable the spring to eject the anti-knock output piston, piston sleeve, output spring and ball.

Take off the rear shell and use a large screwdriver to lever out the non-return valve, following this with the grommet and body gasket.

Turn now to the valve chest, take out the setscrews and remove the lever guide and spring plate (Fig. 6:6). To remove the 'T' lever valve, the plug in the body will probably have to be depressed.

Use a small piece of thick wire or $\frac{1}{8}$ in. welding rod inserted into the hole in the control piston to lever it along and push out the plug (Fig. 6:7). Follow this with the control piston. Compress the spring to remove the circlip, retainer and spring abutment; then remove the seals from the plug and control piston. Dismantling is completed by

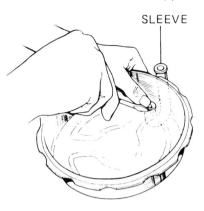

SLEEVE

Fig. 6:4. Removing the sleeve and support plug from its mounting in the shell

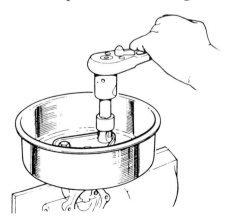

Fig. 6:5. Parting the body (and all its internals) from the rear shell

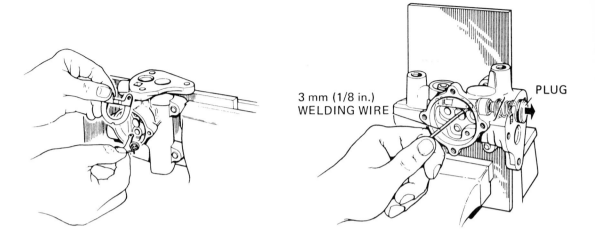

3 mm (1/8 in.)
WELDING WIRE

PLUG

Above **Fig. 6:6. Removing the lever guide and spring plate from the valve chest**

Above right **Fig. 6:7. Levering the control piston along to push out the plug**

removing the air filter and its associated part (Fig. 6:8).

This is where absolute cleanliness becomes vital. Wash your hands and lay out a sheet of clean paper. Check the dismantled parts carefully, looking particularly closely at the control piston and cylinder bores in the hydraulic body for any signs of pitting, corrosion, scoring or wear ridging. If there are any signs at all of poor condition, a new servo unit will have to be obtained.

If these essential parts are in good condition, all the parts from the overhaul kit can be laid out. Checking against these will indicate what to discard and what to retain among the dismantled parts. Clean all the retained parts in Girling cleaning fluid or clean brake fluid, lay them out on the paper and allow to dry. Handle the new diaphragm as little as possible and ensure it always stays clean and dry.

Reassembly begins by lubricating the control piston, piston seals and bore with clean, unused brake fluid. The new seals go on the piston so the lips face away from the centre hole. Fit the spring abutment, spring retainer and circlip to the piston and insert the assembly back into the bore, aligning the piston centre hole with the 'T' lever position in the valve chest. Fit the plug with its new seal and refit into the body.

Install the 'T' lever next, the round end of the lever fitting into the control piston hole. Press the plug until it

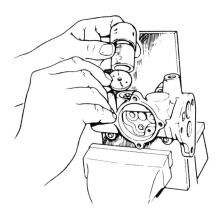 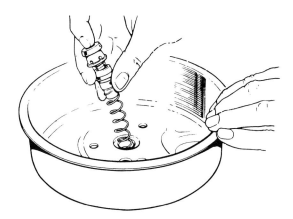

locates properly. Install the spring plate and lever guide in position on the lever, ensuring the slot in the spring plate engages the lug above one of the lever valves. Tighten down all of the retaining setscrews.

Clamp the hydraulic body back in the vice with the piston bores vertical and lubricate the output piston bore and seal with clean brake fluid. Fit the new seal to the anti-knock piston so the lips face towards the reduced end. Position the rear shell half on the body and fit the anti-knock piston, spring and other parts through it into the bore (Fig. 6:9). Take care to ensure the ball remains in the piston and doesn't drop down into the bore. Hand pressure must be used to retain all the parts in the bore until the clamping plate is fitted and secure by bolts and washers (Fig. 6:10).

Into the rear shell, fit the non return valve adaptor/grommet, smear the ribs of the adaptor with the BMS grease in the service kit and push fully into the grommet. Fit the rubber sleeve to the front shell (Fig. 6:11), coat the plug with BMS grease and insert it.

Fit the new diaphragm to the vacuum piston and fit into the front shell. Follow with the big return spring, large end on the diaphragm plate.

Use the support plate again to hold the unit in the vice, as when dismantling, and fit the new retaining band, with bolt

Above left **Fig. 6:8. Removing the air filter and its associated parts**

Above right **Fig. 6:9. Into rebuilding now and here the anti-knock piston and associated components are reassembled into the body**

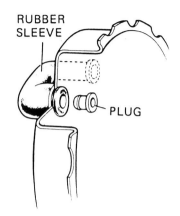

Above **Fig. 6:10. Retaining the parts by hand pressure while the clamping plate is fitted and secured**

Above right **Fig. 6:11. Fitting the sleeve and plug to the front shell**

and nut, loosely on the rear shell. Refit the long vacuum hose on to the non-return valve/adaptor. Fit the gasket and vacuum pipe to the valve chest and secure firmly with the four setscrews and washers.

With the long vacuum pipe connected to the car manifold and the engine switched on, offer the front shell, complete with vacuum piston and diaphragm up to the rear shell, lining up the lines scribed before dismantling. Once the vacuum pipe is pushed into the elbow grommet, the two halves should be held together by the vacuum.

The retaining band can then be positioned, ensuring that the securing bolt and raised legs won't prevent the unit being fitted back on the car in the same position from which it was removed. Ensure also that the little convex V pressings engage with the level rim sections on the front shell.

With the clamping ring in position, gradually tighten the bolt, at the same time tapping each side of the ring with a hammer to ensure the V pressings are correctly positioned on both shells and press them together. When fully tightened, turn up the tabs of the lock washer.

Finally, fit the air filter element and check carefully for any air leaks. Then turn off the engine and refit the unit to the car and bleed the brakes (chapter eleven). The unit may be tested as already described earlier in this chapter.

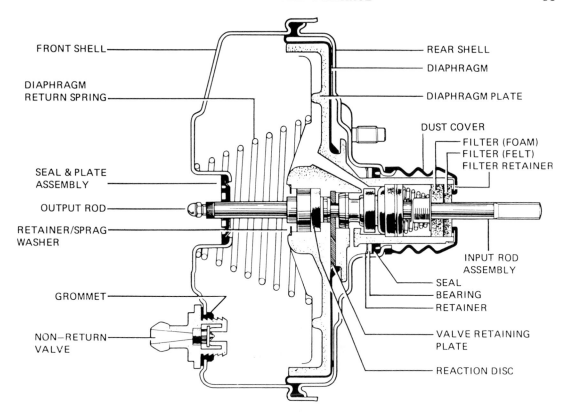

FRONT SHELL

DIAPHRAGM
RETURN SPRING

SEAL & PLATE
ASSEMBLY

OUTPUT ROD

RETAINER/SPRAG
WASHER

GROMMET

NON–RETURN
VALVE

REAR SHELL

DIAPHRAGM

DIAPHRAGM PLATE

DUST COVER

FILTER (FOAM)
FILTER (FELT)
FILTER RETAINER

INPUT ROD
ASSEMBLY

SEAL
BEARING
RETAINER

VALVE RETAINING
PLATE

REACTION DISC

Fig. 6:12. A cutaway side view of the Girling Supervac mechanical servo unit

Mechanical servos

Mechanical servos are normally installed as original equipment and can usually be identified because they have the master cylinder mounted directly on to them. Generally, the possibilities of being able to overhaul a unit of this type are considerably less than with a hydraulic type. Here, the limited servicing possibilities of the Girling Supervac are detailed (Fig. 6:12), a type used in the past by Austin, Morris, Ford, Fiat, Hillman and Vauxhall among others, but similar also in principle to those of other brake manufacturers.

No one will want to go to the expense of changing a servo unnecessarily, so testing to find out whether the unit is working properly would seem to be a good idea.

Raise the front of the car and support it on axle stands, then with the rear wheels chocked, start the engine to build up a vacuum in the servo. Check that one front wheel spins freely, apply the brakes several times, and then re-check that the wheel is still free. If the brakes bind, a major fault is likely.

With the engine running, apply the brake pedal several times. If pedal response is sluggish, suspect that either the vacuum hose or the air filter may need changing. Let the engine run to build up a vacuum, then stop the engine and try the brake action. At least two applications should be power-assisted. The pedal will 'harden' as the vacuum is exhausted. If there is no servo-assistance, the problem could be a faulty non-return valve, blocked vacuum hose, or a leak in the vacuum system.

The last check is to switch the engine off and apply the brakes half a dozen times to exhaust the vacuum. Then, with the foot held lightly on the brake, start the engine again. If the unit is working, the pedal will move down slightly as the vacuum builds up and less pressure is required to keep the brakes applied.

If a major fault is indicated, a replacement unit is the only answer, but three smaller servicing kits are available—filters, non-return valve kit and a service kit.

Because it can be done without removing the servo from the car, the easiest of these service operations is replacing the non-return valve. There are two types—the earlier one, which is more likely to be encountered by the restorer, is a bayonet fixing; the later one is a push fit.

Note the angle of the valve nozzle; making a simple sketch is the most foolproof way of getting it right. Take the vacuum hose off; it may be a push fit or, more likely, a Jubilee clip.

With the bayonet type, the removal method can be seen in Fig. 6:13. You press down on the valve to compress the 'O' ring and then turn it through about 120 deg. with a suitable spanner to release the fixing lugs. Fit the new one together with a new 'O' ring, but fit this dry (no lubricant).

The push-in type is shown in Fig. 6:14 and this is usually pulled out while exerting a side load. Alternatively, it can be levered out with a flat-bladed screwdriver, but taking great care, of course, not to damage the unit. Remove the old grommet, taking care not to drop it inside the vacuum

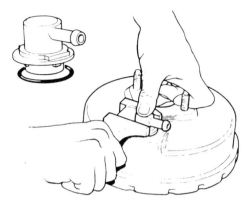

Fig. 6:13. Replacing a bayonet type non-return valve

chamber and then simply fit the replacements, lubricating the ribs of the new non-return valve with the BMS grease in the kit.

Fitting new air filters can sometimes be done with the unit still in the car, but as the vacuum seals will probably be changed at the same time—both jobs are scheduled at 40,000 miles (64,000 km) or three years—and that requires the unit to be removed, it's best to take it out anyway.

To remove the unit, it's best to follow the car manufacturer's instructions, but it's not very complicated, and follows along similar lines to the hydraulic type described earlier.

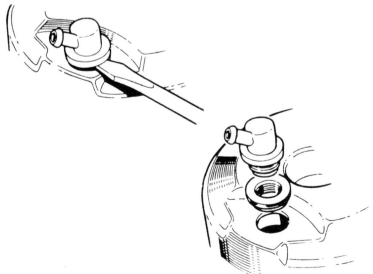

Fig. 6:14. Replacing a push-in type non-return valve and grommet

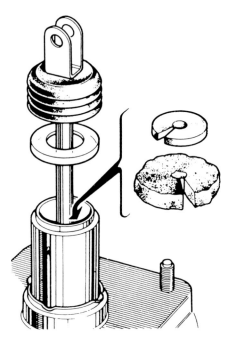

Fig. 6:15. Replacing the air filter and dust cover

Briefly, it's a matter of disconnecting the hydraulic pipes at the master cylinder and plugging the ends to prevent the ingress of dirt. Disconnect the unit from the pedal—usually a simple split pin and clevis and then the servo mounting nuts. The unit will then come off together with the master cylinder. On some cars, it may be possible to leave the master cylinder in situ, which will save disconnecting the hydraulics, although care must be taken not to damage the pipework when manoeuvring the servo out.

The filter assembly which is located at the input rod end of the unit can be seen in Fig. 6:15. To change it simply pull back the rubber dust cover, and take out the filter retainer. Cut the old filter to remove it and the new one to fit it. Arrange the new one to form a complete circle around the rod, refit the retainer and, if you've got a service kit, fit a new dust cover which will be included in it.

Vacuum seals are located between the master cylinder and the servo unit (Fig. 6:16). Take off the master cylinder (two nuts) and put it somewhere clean and safe. Pull out the vacuum seal and retainer using needle-nose pliers, clean and lubricate the walls of the recess (BMS grease from the kit only), and fit the replacements. There are some variations between different Girling models but the work is not generally complicated. If there is a breather hole, ensure it remains unblocked.

When the unit is refitted, the braking system will have to be bled. See chapter 11 for details.

Fig. 6:16. Fitting parts from a Girling service kit, with particular reference to replacing vacuum seals

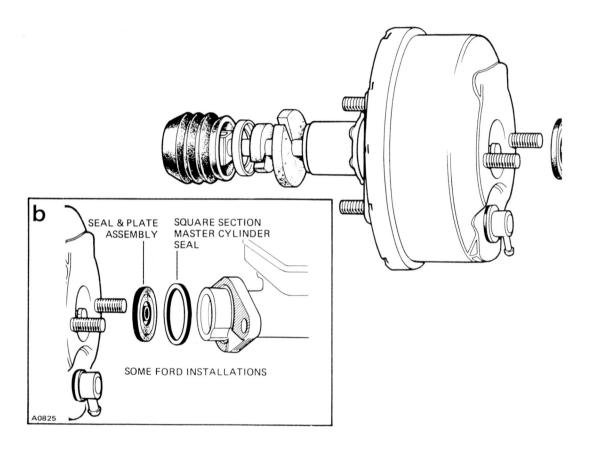

b

SEAL & PLATE ASSEMBLY

SQUARE SECTION MASTER CYLINDER SEAL

SOME FORD INSTALLATIONS

A0825

Chapter 7 | Drums and discs

Drums and discs don't last for ever, but in addition to wearing out, neglect of pads or linings can result in damage. Ignoring this if it's present, and simply changing friction pads and linings, is a waste of time because they will wear out quickly, braking efficiency will continue to be impaired, and it will all have to be sorted out eventually anyway.

Drums

Inspecting drums for damage is usually done when renewing the linings, and it means, of course, taking the drums off. This has been covered in the earlier chapters, but briefly involved jacking up the relevant end of the car, supporting firmly on axle stands, chocking the wheels still on the ground, and removing the road wheels.

Whether the drum is simply held by a couple of countersunk set screws or whether it is a combined drum/hub unit, the next step should be to back off brake shoe adjustment. Removing the countersunk setscrews and clouting with a mallet will usually sort out the former type. The hub/drum unit means prising out the axle grease cap, pulling out a split pin and removing the castellated axle nut and thrust washer. The unit can then usually be pulled off, fishing out the outer wheel bearing at the same time.

If it sticks, refitting the road wheel and using this as a lever, spinning it and clouting round the wheel rim with a mallet, will usually shift it. On some cars a puller is required, notably for something like the rear wheels on a BLMC 1100, where there is also the complication of a left-hand thread on one side.

Fig. 7:1. This is the smooth
surface the disc should present.
The abrasive paper is being used
here to smooth off after
removing rust

Cleaning out a drum is no longer the slap happy use of a
brush or an air line. Now we know that inhaling asbestos
dust is dangerous, a vacuum cleaner or cleaning fluid are
recommended. Use a wire brush on the outside of the drum
and wipe the inside round with a rag dipped in cleaning
fluid.

Now inspect for visible defects; some will be im-
mediately obvious but others may take a bit more
identifying. Principally you'll be looking for deep scoring,
caused by brake shoes left in until the lining material has
gone completely. Deep ridging of this type is often known
as 'tramlining' and a replacement drum is the only answer.
Deep pitting or corrosion mean a new drum too and so does
excessive wear indicated when the swept area is deeply
recessed.

The best method of assessing drum distortion is to
measure it using an internal micrometer. If you don't have
one, you may be able to get one locally from a hire shop.
Measure across the inside of the drum in one direction and
then take a second measurement at right angles to it. The
difference between the two figures will indicate ovality.
The maximum permissible is 0.001 in. per 1 in. (0.010 mm
per 10 mm), which means 0.009 in. (0.229 mm) in a 9 in.

Fig. 7:2. This drum is quite
badly worn, indicated by the
swept area being deeply recessed

(229 mm) drum. Excessive ovality results in snatching, juddery braking and is best avoided.

While using the internal micrometer, take measurements in the other direction as well, measuring first at the closed end, then in the centre of the swept area and then near the rim. This will give an indication of a tapered drum or one that has 'belled'. You will also be able to spot this probably from the way the brake shoes wear. In fact any uneven wear is an indication of trouble, and heavy wear on the ends of the linings could also be an indication of drum ovality.

The swept area on the drum should be a shiny even surface all round and the only damage which is acceptable is minor scratching and scoring, not more than 0.001 in. (0.025 mm) deep.

Undoubtedly the best solution, if a brake drum is worn or damaged, is another drum although some people argue that skimming is an acceptable alternative. Mostly, in Britain this is rejected because it alters the internal diameter compared with that of the shoe, so that the two different shapes do not make good contact. The reduced thickness of metal is also more prone to distortion and does not conduct away heat so efficiently.

Having said that, some manufacturers do accept the

possibility of skimming and actually put a maximum figure on how much metal can be taken off. This is certainly the case on American drum brakes.

Perhaps the most economical way of obtaining a replacement drum is from a breakers yard. Take care, however, to choose one that has none of the faults already listed.

Discs

A disc is easier to check because it can often be checked thoroughly after removing the road wheel. Look first for signs of deep scoring, pitting or cracking. Any of these make a new disc desirable. If there is light rust around the periphery of the clean swept area on the disc, take the opportunity to get rid of it by spinning the disc against the tip of a flat screwdriver. The cause of scoring, of course, is the same as with drum brakes—leaving the pads too long before changing them.

The next thing to check for is excessive run-out on the disc; that is virtually 'wobbling' from side to side between the pads. If this condition exists, you would probably have already noticed it as 'pedal flutter' caused as the disc movement 'knocks back' the pistons.

Checking this will probably mean another visit to the hire shop, this time to pick up a dial gauge. It is set up on a firm base near the disc and the 'probe' adjusted to just touch the disc surface. Slowly rotating this will then show the amount of run-out on the dial. The maximum acceptable figure is 0.004 in. (0.01 mm)—any more than this and a new disc is required.

Similar pedal effects can be felt with a disc that is of uneven thickness. This is checked next using a micrometer at a number of points all round the disc. The variation of thickness should not exceed 0.0005 in. (0.0127 mm); if it does, fit a new disc. Excessive wear, also indicated by a recess under an overhung rim, could be checked out too.

Once again there is the possibility of having the disc ground (not turned) to eliminate damage or to true it up, but this does increase the possibility of distortion and reduce the disc's heat dissipation ability. No more than 0.050 in. (1.27 mm) should ever be taken off the thickness.

Chapter 8 | **Pipes and hoses**

The older a car is the more likelihood there is of trouble with the rigid 'bundy' tubing of the circuit or the flexible hoses. Corrosion is the main problem with the pipework, although there is also a possibility of damage when mounting clips and brackets break. With flexible hoses, the rubber material from which they are made can deteriorate, and does, and there is always the possibility of abrasion wear, though this is usually due to wrong fitting which allow the pipes to rub on steering or suspension components.

Flexible hoses
These are likely to deteriorate and need changing more often than the rigid pipework, so they will be dealt with first. There's not much alternative to visual inspection and the way to do this is to swing the steering onto full lock in each direction so that each hose in turn is as slack as possible. Perishing can usually be spotted by bending the hose into a sharp loop, revealing itself as the typical tracery of fine cracks. Don't just do this in the middle of the pipe, however, this sort of deterioration is just as likely towards the end of the hose, so inspect there too. Look also for 'flats' worn on the hose indicating chafing. Traces of brake fluid are a giveaway too, even when actual leak points cannot be located. As a final test, get someone to operate the brake pedal hard to see if there is any sign of a hose 'ballooning' because it is damaged and weak.

   If you buy your new hoses from a franchised dealer, specifying all the usual vehicle identification numbers, etc., you should have no problem in getting the right hose, but in addition to the usual complications of the right type and the right length, there is also the question of the threads. With older British cars these were all UNF, but in later years metric fitting did start to be used. A quick comparison check, therefore, before dismantling, could save a lot of time and trouble.

There is an accepted method of removing a flexible hose. Start by clamping a sheet of Polythene under the cap of the master cylinder reservoir; this will effectively limit the amount of brake fluid that escapes and make bleeding easier. Refer to Fig. 8:1 and start by undoing the tube nut 'B'. To prevent the hose twisting, hold the hexagon on the hose ('A') with an open-ended spanner and undo the locknut 'C'. The best way is to locate the two spanners with just an inch or two between shafts and then squeeze them together to break the lock between the two nuts. Finally, the hose can be quite easily unscrewed from the brake end.

To be absolutely safe, it is best to plug the ends of the pipes or unions whenever a part is dismantled and only remove the plug when the new component is to be fitted. The copper washer fitted under the clamping nut should be discarded and a new one fitted with the new hose. Reverse the sequence to fit the new hose, tightening the nuts just enough to prevent leakage. The recommended torque figure (just as a guide) is usually around 15 lb ft.

When fitting at the inboard end, pass the hose through the mounting bracket, ensure it is not twisted, and hold it in place with a spanner on the union 'A' while fitting the washer and locknut 'C'. Refit the tube nut, again only tight

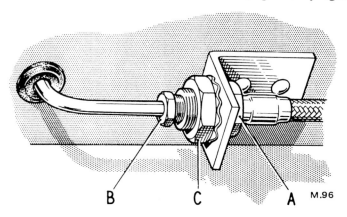

B   C   A   M.96

Fig. 8:1. The method of removing a flexible hose is to undo the tube nut B, then hold A with an open-ended spanner and undo the locknut C

enough to prevent leakage, but before bleeding the system, turn the steering onto both extremes of lock and bounce the suspension to ensure the pipe is clear of anything that could cause chafing. It is generally a good idea to fit new hoses to both sides of the car if one is found to be defective.

Changing a pipe

It is not always too easy to decide whether a pipe is really badly corroded or if there is just a dusting of surface rust; dirt can often disguise the true state of the pipe, so it's a good idea to start by wiping them clean.

A rule of thumb method of checking is to rub the suspect pipe with a coin. If this produces sound bright metal, the pipe is probably sound; if it doesn't and it still looks pitted, the pipe wants changing.

It may be possible to buy the relevant new pipe ready formed and complete with the correct selection of ends and unions, etc. already in place. In this case, the job is relatively easy.

Use a sheet of polythene under the master cylinder reservoir cap to inhibit the escape of fluid, and use the same two-spanner technique already described to undo the unions. When the pipe is off, compare it on the bench with the new one, but before doing this, ensure that you have plugged the ends of the pipes remaining on the car to prevent the ingress of dirt. If the new pipe does not conform to the exact shape of the old one, bend it gently around a suitable former until it's right: then install it.

If you have decided to change the whole braking system, the same technique is used for each section of tubing, tackling them one at a time, starting at the wheel furthest from the master cylinder, and finishing finally with the connections to the master cylinder itself.

Check all the securing clips as you go and if any are missing or break as you take down the old pipe, renew them. One of the modern Girling plastic clips that is fitted like a rivet—you hit it to collapse a section that grips in the mounting hole—is one way. Carefully check also that grommets are in place where a pipe passes through a bulkhead and, if they are in a poor state, renew them anyway. The final job is to top up with new brake fluid and bleed the system through. When all the air is out, a check should be made for leaks by pressing hard on the pedal, whilst checking all the disturbed connections. Once convinced the system is sound, go round and bleed again at every nipple so that the system is totally flushed through and none of the original fluid remains.

A different way of buying replacement brake pipes is from a rack of straight lengths. Girling, for instance, market them this way. Their catalogue details which pipe you want for the job you have in mind and it comes as a straight length, but already equipped with the right unions and fittings. All you have to do is bend the pipe and fit it.

There are proper formers available, but they are hardly worthwhile for occasional use and a whole range of things can be used as substitutes, from a suitably-sized socket to a sewer pipe, depending on the sort of radius you want. Don't be tempted to try shaping without a suitable former; once the pipe is kinked it's useless, and if you try bending it twice on the same place, it will break anyway.

Obviously, you will need to take off the old pipe first and use this as a guide for shaping the new one. It's well worth spending the time and trouble on the bench getting it right; any hassle it saves under the car is a bonus.

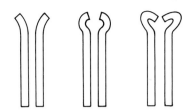

Fig. 8:2. Pipe flares shown diagrammatically. On the left is a single flare. In the centre is the first stage of forming a double flare and the finished shape on the right

Making from scratch

Making up brake pipes is usually regarded as a job for the garage. There are two main reasons for this—first it requires special equipment to make the flares in the ends of the pipes (Fig. 8:2), and second, there's a lot of know-how required to use it. The fact that the cost of a flaring tool kit has come down in recent years makes the work somewhat more of a possibility now for the amateur, and it is also possible sometimes to get a flaring tool kit from a local hire shop. The real professional gear is heavy and bench mounted, but the latest kits available for the amateur use the same principle of operation but employ a threaded clamp instead of operating like a bench vice.

A typical kit uses a special vice with different sized holes in it which will grip various pipes firmly without crushing them while the flaring work is carried out. The initial step is to cut the end of the pipe square, file it clean and remove any burrs both from the inside and the outside. Adaptors from the kit are then used in conjunction with the screw clamp to shape the flare. The single flare shown on the left of Fig. 8:2 is formed by screwing a conical former into the end of it. The more complicated double flare on the right is done by fitting a spigoted adaptor first, the spigot going inside and the main part of the adaptor fitting the outside. Applying force from the threaded clamp squashes the tube into the shape shown in the centre of Fig. 8:2, the spigot preventing it from collapsing inwards. Using the conical adaptor then achieves the final and correct shape by folding the metal in on itself.

When making flares, try to avoid using too much pressure; you need to leave a little more possible compression to enable the tube nut to achieve a good seal when the pipe is fitted. One last thought on this subject, do remember to install all the fittings on the tube before completing the second flare; you can't put them on afterwards!

Pipe flaring can be carried out on either of the materials normally used for brake pipes—the most commonly used thin steel, copper, or the alloy Kunifer 10. Whatever you use, ensure it is purpose-made brake line tubing; nothing else.

The normal type fitted as original equipment is rolled and the seam brazed to form tubing with double walls; it's made of steel, copper coated on the inside and zinc plated on the outside as protection against rust. It's got a high bursting pressure too—16,500 psi compared with an emergency stop braking pressure of only 2,000 psi. It has good fatigue resistance, but it does eventually rust.

Copper tubing is imported from Scandinavia and marketed by Automec in ready-to-bend and fit single pipes and in complete car kits. The pipes are fitted with brass unions, which means they are completely rust-proof and being copper, they are malleable and easy to form.

The main alternative is a seamless, non-ferrous tubing called Kunifer 10, manufactured by Yorkshire Imperial Metals Ltd. It is a copper nickel alloy which will outlast the rest of the car as far as rust is concerned, although it has been said that it is more susceptible to work hardening than

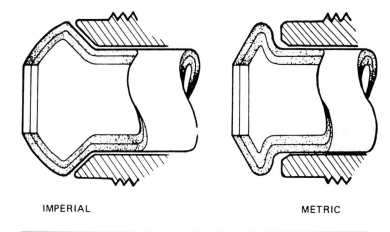

IMPERIAL   METRIC

Fig. 8:3. Sections cut through imperial and metric fittings. Note the difference in shape and the difference in the pipe flares to fit on them

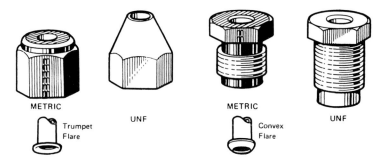

METRIC   UNF   METRIC   UNF
Trumpet   Convex
Flare   Flare

the more commonly used steel.

It is popular with people renewing braking systems on a cherished car they definitely plan to keep; an occasion where the extra cost is definitely justified.

Pipe unions

The majority of restorers, because they are working on older British cars, will be concerned only with UNF fittings, but as there are two types (UNF and metric), and the methods of identifying them have changed, a few comments will be found useful. Figs. 8:3 and 8:4 should be a help initially.

Prior to 1974 all metric pipe nuts, hose ends and bleed screws were coloured black and UNF versions were either silver or gold. Now the metric fittings are coloured pale gold instead of black, and UNF fittings are silver. A letter

Fig. 8:4. The difference between metric and imperial unions. Note that the metric female nut is always used with a trumpet flare (left) and the metric male nut is always used with a convex flared pipe (third from left)

Above **Fig. 8:5.** Some sections of brake pipe are exposed to water and mud thrown up from the road. Like this one, they will eventually rust and need replacing. Similarly, the flexible hoses are also vulnerable

Above right **Fig. 8:6.** Apart from matching the shape and fittings of the old pipe, fixing clips etc. also must be transferred. Check them first and renew them if not satisfactory

Right **Fig. 8:7.** One way of buying replacement brake pipes is ready fitted but straight. Here the new one will have to be shaped to correspond with the old

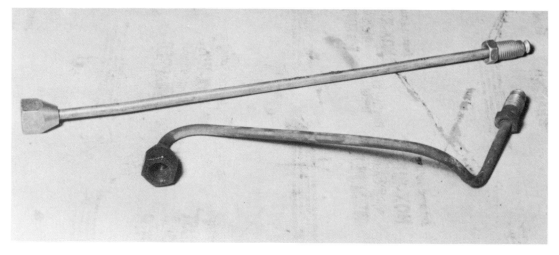

Left **Fig. 8:8. This is the sort of rack you might have to choose from. Selection is done with the aid of a catalogue**

Above **Fig. 8:9. Pipe flaring machinery is not simple, nor is it cheap if it's the professional variety. There are cheaper types available now for the home mechanic**

Fig. 8:10. This is a professional former for bending brake pipes around different radii. Anything with the right sort of curve on it can be used as a substitute

'M' may also be found stamped on many metric nuts.

Most likely to be confused are 10 mm and $\frac{3}{8}$ in. UNF pipe nuts. A $\frac{3}{8}$ in. UNF hose or pipe nut can be screwed into a 10 mm female thread, but it will be very slack and will not seal. The other way round, a 10 mm fitting will screw into a $\frac{3}{8}$ in. UNF female thread, but only just, and will probably jam after a couple of turns. This too, of course, will not seal.

The other fittings are not so likely to cause confusion, but you will need to know whether they are metric of UNF before you order any replacements.

Chapter 9 | Handbrakes

While service brakes are made by just a few specialist brake manufacturers, handbrakes are still the province of the car makers. The result is a wide variety of different design features. Many operate on the rear drum brakes, some using the existing wheel cylinder actuation and others a separate lever system. Some operate on rear disc calipers, either using the existing calipers or separate units; some even have a small drum brake incorporated in the rear disc set-up. Others are totally different with the handbrake working on the front wheels.

Actuation can vary too, ranging from the normal floor lever through 'umbrella' dashboard mounted types, 'flyoff' sporting brakes and even a foot-operated type.

Under the car there are more variations—twin cable systems, a single with a compensator and one that operates on one wheel and is bridged across to the other by a lever. Some are totally rod operated (older cars particularly), some are exposed cables, others are the familiar Bowden type.

Here we will deal with the more common types and touch briefly on the more rarely seen, but some of the general principles apply whatever the mechanism tends to be.

If your car for restoration has been recently acquired, it is not only important that the handbrake should be working, it should also be unlikely to fail in the near future, and also should be capable of passing the MOT test. That means it must hold the car firmly on a gradient and also act as an emergency brake, meeting the MOT retardation figure of 25 per cent.

Checking rods

A visual inspection is a good way to start, and it is useful to have an assistant pull the brake on and off while you watch various sections of the mechanism operate. This will help discover, first whether all the mechanism is operating, second whether anything is broken or seized, and third perhaps whether anything is badly worn.

On a rod system look to see if the rods are bent or damaged or whether they have suffered in the past from rust. A replacement is usually the best bet, but it is possible to take a rod off and straighten it on the bench, checking as you go with a straight edge. Assess also the condition of any clevis pins used, and if they look suspicious or there feels a lot of play in the joint, pull out the split pin, remove the clevis and check it on the bench for wear. If it is heavily grooved and worn, fit a new one and a new split pin.

The actual yokes might also need changing because of holes elongated by wear, and the simplest way to do this is to tackle them one at a time, counting the number of turns necessary to unscrew each rod out. Using the same number when refitting can save a lot of trouble realigning the system afterwards. Once it has been reassembled, coat all the moving parts and the rods with grease as protection against rust, and make sure that the grease gun is used to charge any grease nipples on swivel trees, etc.

Checking cables

There are two types in general use—those which are a simple exposed wire and others which run through an armoured sleeve (Bowden type). Check the former type mainly for damage and fraying; it only needs one strand broken to weaken the cable and make it necessary to fit a replacement. These can also stretch, but before deciding to replace them for this reason, try the effects of adjustment, which is dealt with later in this chapter.

The most likely problems with sheathed cables are seizure, which can be spotted by getting an assistant to operate the handbrake lever, and damage to the outer covering which also usually results in cable seizure.

Replacement is probably the best answer here, but it is sometimes possible to free a seized cable by taking it off the

Fig. 9:1. Typical compensator point where adjustment is also carried out

car, removing any grease nipples, and soaking it in penetrating fluid.

How you take a cable off depends mostly on where in the linkage it is situated. At the front (lever) end, with the lever in the off position, disconnection may involve a clevis pin, or a screwed rod, which merely means undoing the adjustment and locking nuts, or a simple slot-in device, or some other method; it may also be necessary to remove the handbrake lever first.

At the compensator, where this is used, the method of fixing is usually clevis pins. At the brake drums, the fixing may be at the rear of the backplate or it may be inside the drum, but it is likely to be a simple hook-in fixing or another clevis pin.

With sleeved cables, it is probably going to be necessary to take off the whole thing—both inner and outer. There will be clevis pins, or something like that, on the inner cables and clips, clamps, brackets or some sort of threaded mounting on the outer sleeve.

The best sequence for releasing all the fixing points will vary from car to car and in the absence of a workshop manual will have to be found by experiment; once discovered, however, it is a simple matter to reverse it to fit the new cable.

Remember to feed any grease nipples on the cable first before fitting and use all new clevis pins and split pins. An open cable is best lubricated with graphite grease, while a sheathed cable needs feeding with LM (lithium) grease from a good high-pressure grease gun, until new grease oozes from both ends.

Another item which may need to be changed, or perhaps removed to free it if it has seized, is the swivel tree compensator. Disconnecting cables, etc. is likely to be a matter of clevis pins, while the swivel tree itself is probably a screw fixing to the rear axle casing. If the unit has seized, it is best wire brushed and then immersed in penetrating fluid for a few hours. It can then be worked free by twisting it vigorously to and fro. The soaking period is best done with the grease nipple removed, but once it is working, this can be put back and the unit filled with clean LM grease. Once it is refitted the whole outside of the unit can be

Fig. 9:2. A cable frayed like this
should be changed right away.
Apart from being dangerous, the
car would not pass the British
MOT test

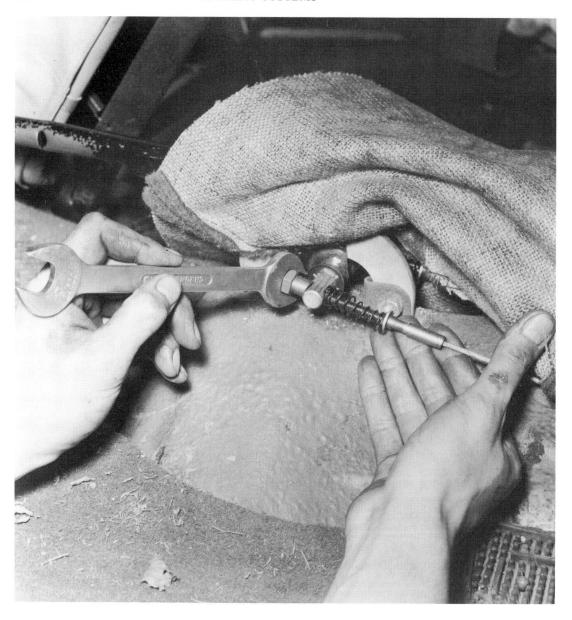

Fig. 9:3. Another form of
adjustment may be found at the
lever end of the handbrake
linkage, inside the car, under the
carpets

coated with grease as a protection against corrosion.

Drum connections

Problems of different sorts can arise at the brake drum end of the handbrake mechanism, and although some of the points have been covered in the drum brake sections of the early chapters, it is worth repeating those likely to be met.

One common system is where both the foot brake and handbrake operate via the wheel cylinder. To spread the braking between the two brake shoes, the cylinder slides in a slot in the backplate. One of the chief problems is that it tends to seize, affecting both the foot brake and handbrake, in that braking is restricted to one shoe only. Ensuring it is free to move and lubricating the contact edges of the slot are part of the normal routine when renewing brake shoes, and may well involve removing the cylinder, cleaning up with abrasive paper and greasing before refitting. Copaslip is an alternative favoured by some people. Extend this lubrication job to the pivots of the handbrake lever.

Another common type of linkage connection inside the drum involves a lever which crosses between the two brake shoes. Seizure of the pivot points can be the problem here and to cure it, the handbrake cable has to be disconnected, the drum removed, followed by the brake shoes, and the linkage cleaned and given a spot of lubricant or Copaslip.

With another type, the cable enters the drum and is connected to a lever, which pivots on one of the brake shoes and is linked across to the other shoe by an operating lever. Free-moving pivots and lubrication are once again the key factors.

Lever problem

One other very important component may need overhauling or possible replacement and that's the actual lever. Normally, it's very robust and as long as it gets an occasional few drops of oil on the ratchet pivot it just goes on working, but nothing lasts for ever.

The most obvious bother is when the handbrake mounting bracket starts to pull away from the floor, with either spot welds or rivets giving up the ghost; the repair, of course, is equally obvious.

Another problem can be when the ratchet and pawl arrangement either starts to stick or slip. When this happens, check with your local dealer's spares counter to find out if parts are available or whether the whole lever has to be purchased. Normally a pawl and ratchet can be obtained, but fitting them means removing the lever from the car anyway. Probably this will only involve a couple of nuts and bolts, but dismantling the ratchet and pawl may be a little more awkward, possibly involving drilling out the rivet-type pivots both the one through the lever itself and the one linking the pawl to the rod connecting to the release button. The new pawl is best attached to the rod using a new rivet, but a suitable bolt with nut and locking nut can form the pawl pivot in the lever.

Removing an 'umbrella' fascia type lever is no more complicated than a floor lever. Disconnect the cable from it first, usually at a clevis fixing, and undo the mounting bolts. Pull out the lever, keeping it horizontal to help retain the pawl and prevent it jumping out until the assembly has been extracted and you can see where the pawl is going. The pawl and ratchet are the parts that wear and which may need replacing. After reassembling, make sure they are both well lubricated.

Adjustment

This may be at several different positions. With two-cable designs it's usually at the lever, with other types it may be at the compensator or at the brake ends of the cable. With many Fords it is at one wheel only, where a single adjuster works for both brake cables. Handbrake lever movement with manually adjusted rear brakes should be only a matter of three or four clicks, but only carry out adjustment after checking every other cause of long travel on the lever. Start with the check right through the system for defective or worn parts and replace them as necessary. Ensure that the rear brakes are correctly adjusted by tightening the manual adjusters until the drums are locked, then backing them off until they are only just free to rotate. A light brushing contact is permissible but they should not be binding at all. If handbrake lever travel is still too great, then embark on the next stage.

Fig. 9:4. Another adjustment point here, this time at the connecting point with the rear drum brake

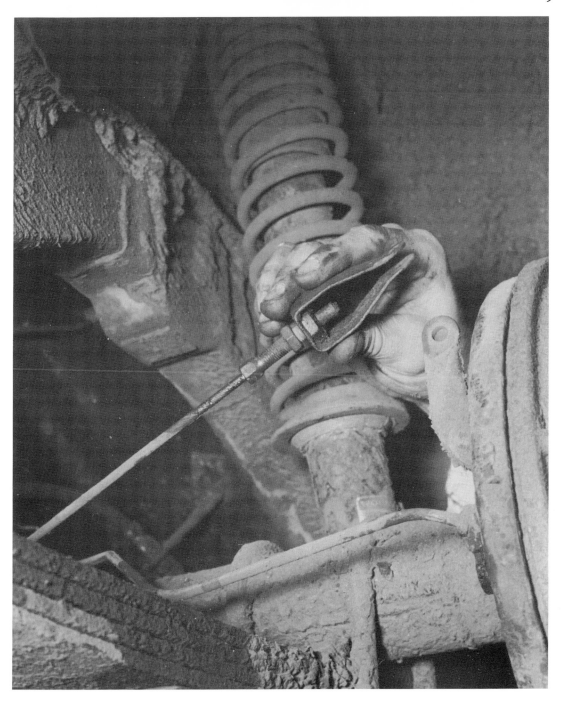

Fig. 9:5. The handbrake cable should be checked along its entire length and its movement watched while someone pulls the lever on and off. The curved guide looks suspect here and even if the cable isn't binding, it should be cleaned and greased

Go back to the rear brake shoe adjusters and tighten them until the drum is locked. Pre-load the cable by pulling the lever on one click. Use the cable adjuster(s) to take up the cable slack and then release the handbrake lever. Re-adjust the rear brakes, slackening the shoes off until the drums are just free to turn. Re-check the handbrake lever travel; it should now be the correct three or four clicks.

With automatically adjusting rear brakes, handbrake travel should be six clicks (although this figure may be much greater on a fuscia type lever if the distances between the ratchet slots are small), instead of the four required with manual adjustment. The reason is that when the drums are hot, they expand in size. The auto-adjuster can then operate to take up the additional clearance. Afterwards, as the drum cools and shrinks, the linings would come into contact with the drum, causing brake drag. Additional clearance is therefore built into the auto-adjuster system to prevent this happening.

Before adjusting to take up excessive handbrake lever travel, make the same check of linkages as for manually adjusted brakes, and replace any worn parts. Additionally, check that the auto-adjust mechanism is working correctly; if not, renew parts as required as detailed in the earlier chapters. Once this has been ascertained, if handbrake travel is still more than eight clicks, adjustment should be carried out as follows.

Disconnect the cable from the clevis on the backplate and apply the handbrake lever half its travel. Reconnect the cable to the clevis and shorten the cable with the adjuster until the brake is applied. Release the handbrake and check that the handbrake lever travel has improved. Check also that, with the handbrake released, there is no drag between shoes and drum; if there is, the cable has been overtightened. Disconnect the cable at the clevis and repeat the above procedure.

If, after adjustment as described, it is not enough to take up all the slack, the cables have stretched and need replacement.

Chapter 10 | Safety testing

The MOT Test was specifically designed to check out the roadworthiness of the older car, but when it comes to the braking system, there is no reason at all why a car of restoration age should be any more difficult to bring up to standard than a current model.

Brakes have improved over the years and so also has the efficiency of the MOT test. Disc brakes and servos were probably the major landmarks in car braking; with the MOT it is the rolling road brake tester. In the early days, a decelerometer was used, but that only measured overall braking, and it was quite possible for it to miss out on all sorts of faults. The Girling company's own research, for instance, showed a very much higher percentage of faults than were ever revealed by the MOT test.

Any dismantling, within the limits of a simple and reasonably priced test routine, could not be contemplated, so whatever was devised had to discover problems without complicated investigation, but it did have to know when things were wrong.

The current machinery does this quite well. It works essentially like a rolling road dynamometer, on which the car is positioned on rollers and drives them against an imposed resistance to simulate road conditions. The brake tester works the other way round and it is the rollers that are driven, thereby turning the wheels of the car. It is against these exterior powered rollers that the brakes are tested. The torque that brake application applies to the rollers is measured by load cells and appears as reading on two large dials.

The rollers at each wheel are driven individually and this

enables each brake's performance to be measured separately. Using different braking techniques involving both foot brake and handbrake, faults can be detected individually on each wheel and a surprisingly accurate indication of possible faults given by the tester. He can pinpoint things like high braking effort, hydraulic leaks, lazy or sticking pistons, hydraulic leaks, distorted discs or oval drums, pads or linings that are oil-contaminated and even where wrong parts have been fitted.

With this sort of 'Big Brother' checking there's not much chance you'll get away with brakes that aren't up to scratch. you don't want to fail the MOT, not only because it means taking the car in twice, but because if you do your own remedial work, it also means paying twice. Check your brakes out first to ensure you don't fail; that's the answer, and it's exactly what the rest of this chapter is all about.

Road test

If you've been driving the car regularly, you might think that if there were anything wrong you'd know, but that doesn't necessarily follow. You can compensate quite unconsciously for a brake pull to one side, for instance, and then one day when you have to stop in a hurry the car slews violently sideways, particularly if the road is wet at the time. It's dangerous, but not only that—the MOT tester will find it, and then it becomes expensive as well.

On your road test then you've got to make an effort to be objective. Find a flat, quiet, empty road and preferably without a steep camber. Drive the car at a steady thirty miles per hour with your hands just clear of the steering wheel, in position but just not grasping it. Then steadily apply the brakes. What you're looking for is brake bias to one side or the other; any sort of steering pull will indicate something that needs investigating.

Pedal action should be firm and the brakes should start to bite with almost no free travel. On a disc/drum system long ineffective movement will probably indicate the need for rear brake adjustment, but with the car that has front drum brakes, these are even more likely to be the cause.

A spongy feel to the pedal is likely to mean air in the system and the need for bleeding, particularly if the pedal

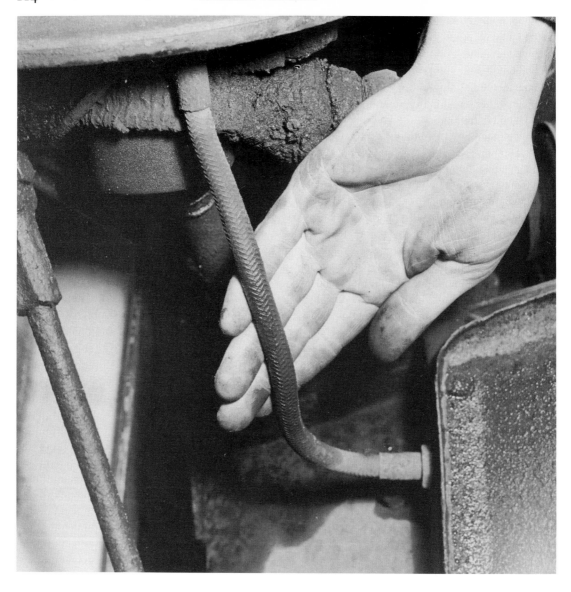

Fig. 10:1. The brake fluid
staining the flexible pipe here is
a dead giveaway. A leaking
flexible could split any time and
result in total braking loss.
Renew it!

hardens up when it is pumped. Snatching or jerking are further indications of problems and so is any kind of noise, particularly the harsh squeal of metal to metal contact.

Try an emergency stop next, a real 'foot on the floor' effort, but do make doubly and trebly sure the road is clear first. If that doesn't produce any problems, pull over to the side of the road and try the handbrake to see if it locks the wheels. Check also that there is not too much lever movement; the MOT inspector will fail it if it takes more than four clicks to lock the wheels with manually adjusted brakes, or more than six/eight clicks with auto-adjust types. One way to check is to try and drive off with it firmly locked on; the engine should stall. One last test—and this is one that the MOT inspector will make himself—push on the brake pedal and hold the pressure to see if it gradually sinks. If it does, it probably means a hydraulic leak—something else you'll have to find if you want to avoid the dreaded 'fail' certificate.

If all these tests produce the right results, you could well be fairly confident of passing the test on brakes, but not necessarily. Brakes that are generally down on performance, for instance, would only be detectable by a driver who has regular experience of other cars. You have to compare overall performance with something else to get any real idea of the standard.

If you found any sort of fault—pulling to one side, snatching, grinding or a long travel pedal, you'll need to investigate this and put the trouble right. You'll find a list of possible causes in out 'Troubleshooter', pages 124/125, and one of the other chapters will say how to rectify it.

Inspection

Curing specific faults is important, of course, but it may not be enough to get the car through the test. What about faults that haven't shown up from the driving check? You really should supplement this by a thorough inspection—the MOT inspector will certainly have a good look round, and he has the advantage that he knows exactly what he's looking for.

Start by checking pad thickness on the front discs; it's only a matter of taking the road wheels off to have a look. If

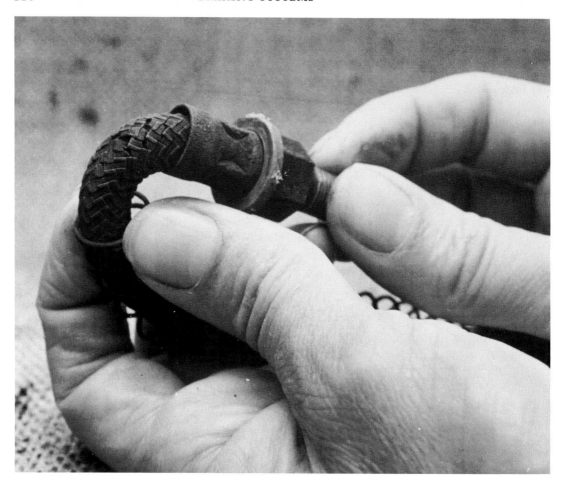

Fig. 10:2. A better way to check flexible pipes is to bend them sharply like this to show up any cracks. This one was taken off a car after an MOT test

you've not inspected the brakes for a long time, it might be an idea to have the pads out so you can see the dust covers on the pistons and check there are no cylinder leaks. Remember too that one pad worn more than the other could mean a seized piston (a very common MOT failure), or in the case of a sliding caliper, seizure between the two parts. Either of these faults should have resulted in brake pull, but you don't always detect it yourself, although the rolling road brake tester certainly will.

Look at the state of the discs, checking for grooving or corrosion damage. Even a slight build-up of rust just outside the swept area is best removed.

Go round the whole system looking for leaks. Check in the vicinity of the master cylinder and the servo unit, if fitted, and trace the pipes from each brake in turn. The most likely sections to rust are often nearest the brakes, particularly if they are exposed to the muck and water thrown up by the road wheels. Scratch any pipe that looks suspicious with a coin; if it remains pitted, it needs replacing. Check the flexible hoses for chafing, perishing or swelling under pedal pressure. Have a look at all the pipe unions and deal with any sign of a leak.

When you get to the back brakes, there's not a great deal you can check without taking the drum off. Even the driving test you made won't have told you much about the back brakes, but you have to remember that the MOT tester is aided by his all-knowing rolling road brake tester, and you don't want him to find something you've missed. Taking the drums off and having a look could be a good idea just in case there's some hidden 'nasty' in there.

The check doesn't have to take long. Look obviously at lining thickness; if they're more than about $\frac{1}{8}$ in. at the thinnest point you're probably OK. Have a look also to see they're wearing evenly, i.e. one piston in the wheel cylinder isn't stuck, or the wheel cylinder is moving in its groove, if that type is fitted.

Any sign of oil on the linings requires immediate action—either a new cylinder if it's that that's leaking, or new axle seals if that's the source of the contamination. Then fit new linings. Have a look under the lip of the cylinder dust cover anyway, even if there aren't more visible signs of leakage. Finally, have a look at the brake drum to ensure it hasn't been chewed up in the past.

That just leaves the handbrake. There's a lot more information on this in the chapter on handbrakes, but briefly, the inspector will check the lever to ensure that the ratchet is sound, even tapping it sideways to ensure it doesn't release itself; he'll test that it locks on and releases properly and will make sure that the lever is firmly mounted (not breaking away from the floor) and that there is no rust in the floor within 12 in. of the lever mounting.

Under the car he'll follow the mechanism from handbrake to brake shoes, checking there are no frayed

Fig. 10:3. This oil-soaked and worn specimen is an ideal candidate for renewal, but don't forget to cure the oil leak first and then change *all* the shoes right across the axle

cables, badly worn clevis fixings, jammed swivel tree compensator, rusty or jammed pulley wheels, or pivoting quadrants, in the case of something like the Mini. Everything should be clean free-moving and well greased. Finally, don't forget adjustment—four clicks at the lever for manually-adjusted brakes and six/eight for auto-adjust types. The automatic adjustment mechanism doesn't have to be working for the MOT, by the way, just so long as the brakes are properly adjusted.

One last thought about the MOT test; it's not just a bit of useless and expensive bureaucracy. It is aimed to ensure that vehicles are properly maintained and comply with legal requirements. But the MOT test and Construction and Use Regulations are only two of the things you risk if your car brakes are anything less than perfect—your life is the third!

Chapter 11 | **Bleed out the air**

The main reason you ever need to bleed the hydraulic brake circuit is to get the air out. Air in the hydraulic fluid is bad news because it is compressible, and depending on how much of it there is in the fluid, it will either result in a spongy pedal or in brakes that don't work at all.

There are two main ways that air gets in—one is because work on the system has meant 'opening' the brake circuit to the atmosphere, and the second is that there is some sort of air ingress in the system. There is actually a third reason for bleeding the brakes and that is when the fluid needs changing; an extended bleeding session is the method that is used.

When working on the braking system it is a good idea to limit the amount of fluid lost and, therefore, the amount of air that is admitted, as much as possible. One way to do this is to clamp a sheet of polythene under the master cylinder reservoir cap. This prevents the entry of air and if air can't get in to replace the fluid, only limited quantities are likely to flow out the other end. Another way to restrict fluid loss, when working on one of the extremities of the system (a caliper or wheel cylinder for instance), is to clamp off the flexible hose, preferably using a clamp designed for the purpose. You could get away with clamping two rods onto the pipe, but *never* use a locking wrench or an ordinary G-clamp directly on the pipe because it will result in damage.

If a spongy brake pedal keeps recurring, you don't need to be a genius to work out that air is getting in somewhere. You might have to be a bit more clever in finding it, however, or very thorough at least. It's a job that has to be done the hard way and that means tracking right through

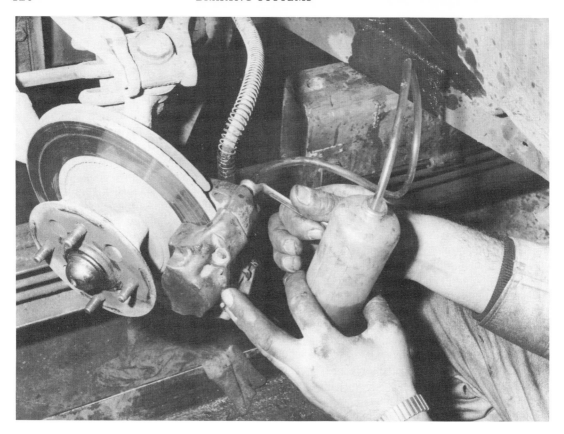

Fig. 11:1. These are the items needed for bleeding air out of the braking system. The spanner must be a good fit on the bleed nipple and so must the polythene pipe. A glass (non-food) container is best to hold the clean fluid

the system looking for the leak. Find it, and sort it out first, or you're wasting your time bleeding the system again.

Changing the entire fluid content of the system is something that is recommended by the brake companies and the brake fluid makers, and although different intervals are recommended by different people, the safest is undoubtedly the shortest, which is every 18 months or every 18,000 miles, whichever comes soonest. The reason is because of a phenomenon called 'vapour lock'. Brake fluid is hygroscopic and absorbs water. Water lodged near the business end of disc brakes, for instance, gets hot and can boil. If it does, all the braking goes. People who have experienced this tend to be extremely careful about their brake fluid!

The equipment required for bleeding brakes is not very

complicated. You need a glass jar, but not one with any connections with the food and drink industry, because it might get back into circulation, and brake fluid is poisonous. You need a transparent bleed tube that is a tight fit over the bleed nipple, a ring spanner for the bleed nipple, a can of new brake fluid, and an assistant to pump the brake pedal.

At one time there was a great deal of controversy about brake fluid. The brake manufacturers said, 'Use nothing but our own brand or all the rubber seals will swell.' Other people maintained there was no difference and they could all be mixed. Now the answer seems to be that the well known and reputable brake fluids, provided they conform to certain standards, are miscible, but there's little point in taking the risk when it's easy to decide on one reputable make and stick to it. When buying incidentally, the earlier standards were SAE 70R3 and SAE J70C; nowadays the high standards that fluid must meet are SAE J1703 and FMVSS116 DOT3. There are two important points. First, use the correct high standard fluid, and second, ensure it is unused and fresh. Aerated fluid that is bled out of the system should be thrown away to prevent it being used again by accident.

The sequence in which the bleed nipples are tackled depends on the brake system fitted to the car. If it's an all-drum or an all-disc system, start with the wheel farthest from the master cylinder and work back to the nearest one. If it is a disc-front, drum-rear combination, do the discs first, followed by the drums, but still bleeding the furthest one of each pair first.

Remove the master cylinder reservoir cap and top up with new fluid. Refit the cap. Then, under the car, push the plastic bleed tube onto the bleed nipple, thread the spanner up onto the nipple and dangle the other end of the tube into an inch or so of clean new brake fluid in the jar.

Undo the nipple a half turn and get your assistant to pump the brake pedal up and down a few times, stopping with the pedal flat on the floor. At this point lock up the nipple and check that the fluid level in the master cylinder is not getting low. Top it up and repeat the pumping action again, watching the dirty-looking gungy fluid and bubbles being ejected into the jar. Directly the bubbles stop, halt the pumping sequence with the pedal on the floor and tighten the bleed nipple.

Repeat this sequence around all four brakes, ensuring that the fluid level in the master cylinder at no time drops to the bottom. If it does, you'll suck more air in and have to start all over again.

Just in case the assistant seems to have it too easy, merely whacking the pedal up and down, there are some complications. The recommended Girling pumping action for their widely used CV master cylinder is to push the pedal down through a quick full stroke, followed by three short strokes, and then removing the foot, allowing it to spring up. Repeat this sequence until all the air is out and then hold it down on the last stroke while the nipple is closed.

With the Girling CB cylinder (this has a cast-iron body instead of the more normal alloy body on the CV), a slow and deliberate push and return action is used, followed by a several second pause, then the sequence is repeated.

On Girling systems where a G-valve is fitted (the BLMC 1800 and some Peugeots have this), put the handbrake on first. Bleed the front brakes according to the master cylinder fitted and then use the CB sequence for the rear brakes. If the pedal suddenly goes hard, the ball in the G-valve has been actuated. Close the nipple and release the handbrake, which will push the ball of its seat and allow the fluid back through. Then continue bleeding.

With Lockheed brakes, the pedal action recommended is to push the pedal down slowly and allow it to spring back unassisted.

When it is proposed to change the fluid, simply extend the bleeding operation until clean new fluid is ejected at each bleed nipple. Some people advocate flushing the system through as well, and here the technique is to fill the master cylinder with methylated spirits instead of fluid and bleed this through until it flows through every bleed nipple. Then change back to brake fluid and repeat the process until pure brake fluid is ejected at every nipple. Make absolutely sure, however, that all the methylated spirit is finally ejected before driving the car.

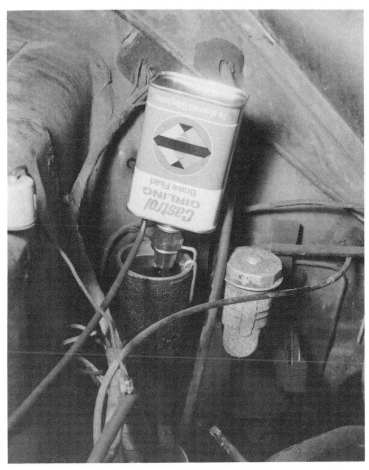

To avoid the necessity of an assistant, there are devices on the market incorporating a one-way valve. One is a set of ABV Automatic Bleed Valves, which fit in place of the normal bleed valves, and the other is the Vizibleed, which incorporates a one-way valve in the plastic bleed tube. With either of these you can nip down under the car to set things up and then go back topside and do your own pedal bashing.

Once bleeding is finished, always work the pedal hard a few times and then re-check the bleed nipples to make sure none is leaking. Leave the system to 'rest' for a few minutes and then try the pedal to ensure it is not spongy; if it is, you'll have to go back and have another go.

Fig. 11:2. At the other end, the master cylinder reservoir must be kept topped up and you can either do it manually, or this handy little device will save you the trouble

Note: One very important point. If you have just fitted new pads and/or shoes, the pedal may feel spongy until the lining material has bedded properly.

Troubleshooter

SYMPTOM	CAUSE	CURE
Excessive pedal travel	**Disc brakes** Excessive hub end float; discs running out; distorted damping shims; wrong master cylinder or too much clearance on pushrod.	Adjust bearings; renew disc(s); new pads and shims; change master cylinder.
	Drum brakes Brakes need adjustment or re-lining; master cylinder vent blocked.	Adjust drum brakes or renew linings; unblock master cylinder vent.
Spongy pedal	Air in system; faulty brake shoes or shoes not run in; master cylinder loose on mounting.	Bleed the system; fit new brake shoes if faulty or allow new shoes to run in; tighten master cylinder mounting bolts.
Pedal requires pumping in order for brakes to operate	Air in system; master cylinder loose; master cylinder defective; drum brakes need adjusting	Bleed the system; tighten master cylinder mounting bolts; renew master cylinder.
Brakes judder or squeal	Distorted drums or discs; loose drum backplate; linings picking up; grease or oil on linings; worn discs; wrong pad material; damping shims or spring omitted; linings incorrectly fitted.	Fit new drums or discs; tighten backplate mountings; chamfer or fit new linings; fit new discs; fit new pads; fit shims or springs.
Brakes 'grab'	Linings picking up; distorted drums or discs.	Chamfer linings or fit new; fit new drums or discs.
Vibration or 'pulsing' on pedal	Drums cracked or distorted; discs out of true.	Fit new drums or discs.

kes pull to one side.	Uneven tyre pressures; suspension or steering faults; seized pistons; oil on linings (one side); uneven adjustment; heavily worn linings; distorted drum or disc.	Inflate tyres correctly; check steering, suspension and dampers, overhaul or replace caliper pistons or wheel cylinders; fit new pads or linings and cure oil leak; readjust brakes; replace linings, drums or discs.
kes drag or fail to release.	Shoes binding on drums (over-adjusted); pull off springs weak; wheel cylinder seized; handbrake cables seized; blocked air hole in master cylinder reservoir; master cylinder fault; servo defective	Readjust drum brakes; fit new pull-off springs; fit new wheel cylinder; fit new handbrake cables; clear master cylinder air vent; fit new master cylinder; overhaul or fit new servo unit.
rd pedal/inefficient braking.	Incorrect, worn, or oil-soaked linings; badly scored discs or drums; seized hydraulics; wrong master cylinder; servo unit not working.	Fit new linings; fit new discs or drums; overhaul or change caliper pistons or wheel cylinders; fit correct master cylinder; overhaul or change servo unit.
kes fail suddenly (no pedal).	Master cylinder failure; burst flexible hose or bundy pipe or union; vapourization of brake fluid (vapour lock).	Overhaul or fit new master cylinder; fit new flexible hoses; fit new pipes, etc; change the brake fluid completely.

Index